AF555917

AGRONOMY
Principles and Practices

About the Authors

Dr. E. Somasundaram is the Professor of Agronomy and Head, Depertment of Sustainable Organic Agriculture, Tamil Nadu Agricultural University, Coimbatore. He completed his B.Sc.(Ag.) at Annamali University with distinction. He did his M.Sc (Ag.) and Ph.D. in Agronomy with specialization in organic farming at AC&RI of TNAU, Coimbatore. He passed M.Sc(Ag.) with University rank and won S.R.P.P. Chettiar Gold Medal and Shield, A.L. Mudaliar and Rao Saheb Murugesa Mudaliar awards for the best M.Sc(Ag.) Student of the University. He also won PPIC Gold Medal for the best M.Sc(Ag.) Thesis, and IOB Gold medal for the best M.Sc (Ag.) Student in Agronomy. He was awarded the ICAR Junior Fellowship for pursuing M.Sc (Ag.). He was a recipient of the Jawaharlal Nehru Memorial Fund award for the year 1992 for his academic excellence. He Passed Ph.D., with Prof.Subramanian and Dr.K.K.Subbiah awards for the best student in Agronomy. He also received T.Konda Reddy award for Agronomy from Madras Agricultural Students Union and TNAU Best Researcher award. As a leader of the team received the best performing tribal subplan centre award during 2014 and best performing OFR centre award for DICRP-IFS, TNAU during 2015 and also best performing ICAR-NPOF centre award during 2016. He is teaching undergraduate students for the past twenty years and has 60 research publications both in International and National Journals and several popular articles to his credit. He continues to teach UG, PG and Ph.D and has so far guided 2 Ph.D and 4 MSc Ag students. He has completed many externally funded research projects.

Dr. M. Mohamed Amanullah is Professor of Agronomy in the Department of Agronomy, Tamil Nadu Agricultural University, Coimbatore. He completed his B.Sc.(Ag.) at AC & RI, Madurai. He did his M.Sc(Ag.) and Ph.D in Agronomy with specialization in farming systems at AC & RI, TNAU, Coimbatore. His teaching experience spans for 27 years to UG, PG and Ph.D degree programmes. He has guided eight M.Sc. (Ag) students and 3 Ph.D students. He was associated with forage crops since his postgraduation and currently with maize crop. His research interest ranges from nutrient management to farming systems. He has completed three research projects. He has edited 7 books, published more than 160 research articles both in international and national journals apart from many popular articles to his credit.

AGRONOMY

Principles and Practices

E. Somasundaram
Professor of Agronomy
Head, Department of Sustainable Organic Agriculture
Agricultural College and Research Institute
Tamil Nadu Agricultural University
Coimbatore

M. Mohamed Amanullah
Professor of Agronomy
Department of Agronomy
Agricultural College and Research Institute
Tamil Nadu Agricultural University
Coimbatore

NEW INDIA PUBLISHING AGENCY
New Delhi – 110 034

NEW INDIA PUBLISHING AGENCY
101, Vikas Surya Plaza, CU Block, LSC Market
Pitam Pura, New Delhi 110 034, India
Phone: + 91 (11)27 34 17 17 Fax: + 91(11) 27 34 16 16
Email: info@nipabooks.com
Web: www.nipabooks.com

Feedback at feedbacks@nipabooks.com

ISBN : 978-93-85516-74-0

Composed, Designed by NIPA

TAMIL NADU AGRICULTURE UNIVERSITY
Coimbatore - 641 003
Tamil Nadu, India

Dr. K. Ramasamy, Ph.D.
Vice-Chancellor

Foreword

Agriculture has been a mainstay of human being since time immemorial. Any country seeking to develop its economy has to give significant priority to agriculture particularly scientific agriculture. It is obvious that, wherever scientific methods are applied, the agriculture got revolutionized.

In agriculture, agronomy is a division that deals essentially with all aspect of soil, crop and water management to increase the production potentiality of crops. It is, therefore, essential that the students of agriculture should have a good knowledge on the basic principles of agronomy.

Presently, textbooks on the basic concepts of agronomy suitable to meet the subject needs for undergraduate students are very few. The book on "Agronomy: Principles and Practices" by Dr. E. Somasundram and Dr. M. Mohamed Amanullah is a timely and welcome attempt to fill this void.

The authors have considerable experience in teaching and research in agronomy.

In this book an effort has been made to gather and compile important informations needed for easy understanding of the various principles of agronomy at the entry level of students in U.G. Programme. I hope thaqt this book will be of immense use to undergraduate students studying Agronomy.

I congratulate the authors for the hard work they put in bringing out this text book in time.

(K. Ramasamy)

Place : Coimbatore - 641 003
Date : 26.09-.2014

Preface

Understanding the basic principles of agronomy is as much important as that of knowing the latest developments scenario in the field of agriculture. It is strenuous strive to keep pace with the progress of such a vast subject like agronomy which is in practice throughout the globe. The budding agronomist has to brace himself with the fundamentals of the agronomy right from the first year of B.Sc (Agri). Under this circumstance, important and relevant information about the various principles and practices of Agronomy are compiled in a book form entitled "Agronomy: Principles and Practices". The book is divided into 15 chapters and which covers comprehensively the syllabus of the principles of agronomy offered for first B.Sc (Agri) as stepping stone to other courses in agriculture particularly the rest of the agronomic courses.

The authors acknowledge their indebtedness to the authors/ publishers of various books from which they have drawn the matter for compiling this book. The list of which is attached at the end.

The authors thank Dr. K. Ramasamy, Vice-Chancellor, Tamil Nadu Agricultural University for providing foreword and necessary guidelines for bringing out this book. The authors profusely thank all the Professors of the Department of Agronomy, TNAU, Coimbatore, for extending useful suggestions for drafting this book.

The authors are much indebted to the Tamil Nadu Agricultural University, Coimbatore for providing an excellent opportunity to bring out this piece of work.

The authors thank their family members for their help, physical and moral support and M/s. New India Publishing Agency (NIPA), New Delhi for their sincere efforts in bringing out this book in time.

E. Somasundaram
M. Mohamed Amanullah

Contents

Foreword .... *v*
Preface .... *vii*

1. Agriculture **1**
Definition 1
Importance of Agriculture 2
Scope of Agriculture in India and Tamilnadu 3
Revolutions in Agriculture 4
Land Utilization Statistics (2012-13) 5
Branches of Agriculture 5
Development of Scientific Agriculture in the World and India 8
National Institutions for Agricultural Research 15
Indian Agriculture and Economy 18
Women in Agriculture and Empowerment 21
Questions 23

2. Agricultural Heritage in India **25**
Agricultural Heritage in India 25
Chronology of Agricultural Technology Development in India 32
Agriculture in Arthasasthra 39
Agriculture in the Sangam Literature 40
Rainfall Prediction 44
Questions 47

3. Agronomy and Agroclimatic Zones ... 49
Definitions ... 49
Scope of Agronomy ... 49
Dimensions of Agronomy ... 51
Relation of Agronomy to Other Sciences ... 52
Role of Agronomist ... 52
Agro-climatic Zones ... 53
Agro Ecological Zones of India ... 61
Questions ... 63

4. Crops and Cropping Sysem ... 65
Classification of Crops ... 65
Crop Adaptation and Distribution ... 74
Theories Governing Crop Adaptation and Distribution ... 75
Major Crops of India ... 75
Factors Governing Choice of Crops and Varieties ... 80
Cropping Pattern and Cropping Systems ... 82
Intensive Cropping ... 83
Crop Rotation - Principles and Advantages ... 85
Factors Determining the Cropping System of a Locality or Region ... 87
Factors Determining the Cropping Pattern ... 88
Cropping Pattern Followed by India ... 89
Questions ... 91

5. Factors Affecting Crop Production ... 93
Internal Factors ... 93
External (Environmental Factors) ... 94
Effects of Temperature on Crop Production ... 97
Questions ... 108

6. Soils ... 111
Classification Based on Soil Taxonomy ... 112
Major Soils of India ... 112
Questions ... 118

7. Seasons ... 121
Agricultural Seasons ... 121
Characteristics of Seasions ... 122

Effect of Season on Choice of Crops 124
Questions *124*

8. Systems of Farming 125
Comparision of Some Selected Features of Different Farming Systems 126
Sustainable Agriculture 127
Integrated Farming System (IFS) 133
Possible Enterprises 134
Dryland Agriculture 135
Questions *138*

9. Tillage and Tilth 141
Definition 141
Characteristics of Good Tilth 141
Objectives of Tillage 142
Types of Tillages 143
Factors Affecting (Intensity and Depth of) the Tillage Operations 146
Depth of Ploughing 147
Modern Concepts in Tillage 148
Tillage Implements 152
Primary Tillage Implements 152
Secondary Tillage Implements 159
Inter Cultural Implements 163
Special Purpose Implements 164
Land Leveling Implements 165
Land Shaping Implemenets 166
Sowing Implements 167
Implements for Wetlands 168
Questions *169*

10. Seeds and Sowing 171
Importance 171
Characteristics of A Good Quality Seed 171
Seed Treatment 172
Objectives 172
Methods of Seed Treatment 173
Sowing 173

Germination 178
Questions *181*

11. Plant Population and Crop Geometry 183
Crop Stand Establishment 183
Optimum Plant Population 183
Importance of Plant Population 183
Factors Affecting Plant Population 184
Plant Geometry 186
Inter Cultivation 188
Questions *189*

12. Weeds 191
Definition 191
Characteristics of Weeds 191
Harmful Effects of Weeds 191
Beneficial Effects of Weeds 192
Crop Weed Competition 193
Methods of Weed Control 194
Integrated Weed Management 199
Questions *200*

13. Irrigation 203
Time of Irrigation 203
Irrigation Methods 204
Surface Method of Irrigation 204
Sub-surface Irrigation 208
Pressurized or Modern Irrigation Systems 210
Water Requirement of Crop 212
Factors Influencing the Water Requirement of Crops 213
Critical of Periods of Water Requirement for Various Crops 214
Questions *216*

14. Manures and Fertlizers 219
Classification of Essential Elements 220
Role of Nutrients in Crop Production 221
Nutrient Deficiency Symptoms of Crops 225
Manures 225

Fertilisers 226
Role of Manures and Fertilisers 226
Agronomic Interventions Forenhancing FUE 227
Classification of Manures 229
Classification of Fertilisers 236
Factors Affecting Manures and Fertilizers Use 242
Time of Application of Manures and Fertilizers 244
Method of Application of Manures and Fertilizers 244
Integrated Nutrient Management (INM) 246
Questions *247*

15. Harvesting and Post Harvest Technology 249
External Symtoms of Physiological Maturity 250
Harvest Maturity Symptoms 251
Criteria for Harvesting of Crops 251
Methods of Harvesting 252
Post Harvest Processing 254
Post Harvest Technology 255
Questions *256*

Selected References 259

Annexures 261
Annexure - I : Units Related to Crop Production 261
Annexure - II : Area (2011-12) 263
Annexure - III : List of Crops: Common and Botanical Names 265
Annexure - IV : List of Major Weeds in the World 270
Annexure - V : Twenty Six Most Common Weeds in Crop Fields of India 271
Annexure - VI : Contribution of Agriculture to National Income . 272
Annexure - VII : Food Grains Production in India - A Scenario .. 273

1

Agriculture

Agriculture is derived from two Latin words **Ager** and **Cultura**. **Ager** means land or field and **Cultura** means cultivation. Therefore the meaning of the word agriculture is cultivation of land.

Definitions

Agriculture is defined in the Agriculture Act, 1947, as "including horticulture, fruit production, seed production, dairy farming and livestock breeding/keeping, the use of land as grazing land, meadow land, osier land, market gardens and nursery grounds, and the use of land for woodlands where that use is ancillary to the use of land for other agricultural purposes".

Agriculture is defined as 'purposeful work through which elements in nature are harnessed to produce plants and animals to meet the human needs'. It is a biological production process which depends on the growth and development of selected plants and animals within the local environment.

Agriculture is defined as **the art, the science and the business of producing crops and the livestock for economic purposes.**

As an art it embraces knowledge of the way to perform the operations of the farm in a skillful manner, but does not necessarily include an understanding of the principles underlying the farm practices. The skill is categorized as;

Physical skill: It involves the ability and capacity to plough the field in an efficient and quick manner, handling of farm implements, animals, sowing of seeds, fertilizer and pesticides application etc., For example: Taking a straight furrow.

Mental skill: It involves the ability of the farmer to take a decision based on his experience, he decides the correct and accurate timing of operation, such as i) time and method of ploughing ii) selection of crop and cropping system to suit soil and climate, iii) adoption of improved farm practices.

As a science it utilizes all technologies developed based on scientific principles such as plant breeding, crop production techniques, crop protection, economics etc. to maximize the yield and profit. For example, new crops and varieties developed by hybridization, development of fertilizer responsive varieties, transgenic crop varieties resistant to pests and diseases, Soil, crop and water management, herbicides to control weeds, use of bio-control agents to combat pest and diseases etc.,

As the business: As long as agriculture is the way of life of the rural population production is ultimately bound to consumption. But agriculture as a business aims at maximum net return through the management of land, labour, water and capital, employing the knowledge of various sciences for production of food, feed, fibre and fuel. In recent years, agriculture is commercialized to run as a business through mechanization.

Importance of Agriculture

Indian economy is agriculture oriented economy. Agriculture helps to meet the basic needs of human and their civilization by providing food, clothing, shelters, medicine and recreation.

Hence, agriculture is the most important enterprise in the world. In a true sense, it is a productive unit where the free gifts of nature namely land, light, air, temperature and rain water etc., are integrated into single primary unit indispensable for human beings. Secondary productive units namely animals including livestock, birds and insects, feed on these primary units and provide concentrated products such as meat, milk, wool, hide, eggs, honey, silk and lac.

Agriculture provides food, feed, fibre, fuel, furniture, raw materials and feed back materials for and from factories, funds, flood control, a free, fare and fresh environment, abundant food for driving out famine, favours friendship by eliminating fights.

Satisfactory agricultural production brings peace, prosperity, harmony, health and wealth to individuals of a nation by driving away distrust, discord and anarchy. It helps to elevate the community consisting of different castes and clauses, thus it leads to a better social, cultural, political and economical life.

Agricultural development is multi directional having galloping speed and rapid spread with respect to time and space. After green revolution, farmers started using improved cultural practices and agricultural inputs in intensive cropping systems with labour intensive programmes to enhance the production potential per unit land, time and input. Agronomist (Scientist dealing with field crop production and soil management) provide suitable environment to all these improved genotypes to foster and manifest their yield potential in newer areas and seasons. Agriculture consists of growing plants and rearing animals in order to yield produce and thus it helps to maintain a biological equilibrium in nature.

Scope of Agriculture in India and Tamilnadu

- Population pressure is increasing but area under cultivation is static or even shrinking, which demands intensification of cropping and allied activities in two dimensions i.e. time dimension and space dimension.
- India is endowed with tropical climate with solar energy throughout the year, which favours growing crops throughout the year.
- There is vast scope to increase irrigation potential in India by river projects and minor irrigation projects.
- India is blessed with more labour availability.

Agriculture is the primary sector. Other sectors are dependent or agriculture. In India major allocation has been done in each five year plan to agriculture. With a 16% contribution to the gross domestic product (GDP), agriculture still provides livelihood support to about two-thirds of country's population. The sector also provides employment to 58% of country's work force and is the single largest private sector occupation.

Agriculture accounts for about 15% of the total export earnings and provides raw material to a large number of Industries (textiles, silk, sugar, rice, flour mills, milk products).

Rural areas are the biggest markets for low-priced and middle-priced consumer goods, including consumer durables and rural domestic savings are an important source of resource mobilization.

The agriculture sector acts as a wall in maintaining food security and in the process, national security as well. The allied sectors like horticulture, animal husbandry, dairy and fisheries have an important role in improving the overall economic conditions and health and nutrition of the rural masses.

To maintain the ecological balance, there is need for sustainable and balanced development of agriculture and allied sectors.

The eyes and minds are soothed by dynamic changes from brown (bare soil) to green (growing crop) to golden (mature crop) and bumper harvests in Agriculture.

Agriculture helps to elevate the community consisting of different castes and communities to a better social, cultural, political and economical life. Agriculture maintains a biological equilibrium in nature. Satisfactory agricultural production brings peace, prosperity, harmony, health and wealth to individuals of a nation by driving away distrust, discord and anarchy.

Revolutions in Agriculture

- Through white revolution, milk production quadrupled from 17 million tonnes at independence to 69 million tonnes during 1997-98 and multiplied many fold now (140 million tones in 2013-14).
- Through blue revolution, fish production rose from 0.75 million tonnes to 5.0 million tonnes during the last five decades and now nearly 13 million tones in 2015-16.
- Through yellow revolution oil seed production increased 5 times (from 5 million tonnes to 25 million tonnes) since independence and cureently 32.88 million tones (2013-14).
- Similarly, the egg production increased from 2 billion at independence to 73.9 billion, sugarcane production from 57 million tonnes to 350 million tonnes, cotton production from 3 million bales to 39.8 million bales during 2013-14, which shows our sign of progress.
- India is the largest producer of fruits and second largest producer of milk and vegetable globally.
- Presently the situation is from food scarcity to food security.

In future, agriculture development in India would be guided not only by the compulsion of improving food and nutritional security, but also by the concerns for environmental protection, sustainability and profitability. By following the General Agreement on Trade and Tariff (GATT) and the liberalization process, globalization of markets would call for competitiveness and efficiency of agricultural production.

Land Utilization Statistics (2012-13)

India

1.	Total geographical area	:	328.726 million ha.
2.	Total reporting area	:	305.936 million ha.
3.	Area under cultivation (Net Area Sown)	:	139.932 million ha.
4.	Total cropped area	:	194.399 million ha.
5.	Area sown more than once	:	54.467 million ha.
6.	Area not available for cultivation	:	43.738 million ha.
7.	Area under forest	:	70.007 million ha.
8.	Other uncultivated land excluding fallow land	:	25.976 million ha.
9.	Fallow lands	:	26.283 million ha.

Tamilnadu (2012-13)

1.	Total geographical area	:	13.06 million ha.
2.	Total reporting area	:	13.03 million ha.
3.	Net area sown	:	4.54 million ha.
4.	Total cropped area	:	5.14 million ha.
5.	Area sown more than once	:	0.90 million ha.
6.	Land put in to non-agri uses	:	2.18 million ha.
7.	Area under forest	:	2.13 million ha.
8.	Current fallow lands	:	1.31 million ha.
9.	Other fallow lands	:	1.70 million ha.
9.	Barren and unculturable land	:	0.49 million ha.
10.	Culturable waste land	:	0.33 million ha.
11.	Permanent pastures & Grazing lands	:	0.11 million ha.
12.	Land under Misc.tree crops and grooves not included in the net area sown	:	0.25 million ha.

(*Source* : Directorate of Economics and Statistics, Govt. of India)

Branches of Agriculture

Agriculture has 3 main spheres namely Geoponic (Cultivation of earth), Aeroponic (cultivation in air) and Hydroponic (cultivation in water).

Branches of Agriculture: Agriculture is the branch of applied science encompassing the applied aspects of basic sciences like life sciences, chemistry and physics. Agriculture includes a wide range of sub-branches like

1. Agronomy : Study of field crops and their management including soil management. It deals with the production of various crops which includes food crops, fodder crops, fibre crops, sugar, oilseeds, etc. The aim is to have better food production and how to control the diseases
2. Horticulture : Deals with the production of fruits, vegetables, flowers, ornamental plants, spices, condiments and beverages.
3. Forestry : Deals with production of large scale cultivation of perennial trees for supplying wood, timber, rubber, etc. and also raw materials for industries.
4. Animal sciences : Deals with agricultural practice of breeding and raising livestock in order to provide food for humans and to provide power (draught) and manure for crops.
5. Fisheries : Deals with practice of breeding and rearing fishes including marine and inland fishes, shrimps, prawns etc. in order to provide food, feed and manure.
6. Agrl. Engineering : Deals with farm machinery for field preparation, inter-cultivation, harvesting and post harvest processing including soil and water conservation engineering and bio-energy
7. Home Science : Deals with application and utilization of agricultural produces in a better manner in order to provide nutritional security, including value addition and food preparation.

On integration, all the seven branches, first three is grouped as for crop production group and next two animal management and last two allied agriculture branches.

Agriculture in practice is broadly grouped in four major categories as follows :

A. Crop Improvement (i) Plant breeding and genetics
(ii) Bio-technology

B. Crop Management (i) Agronomy
(ii) Soil Science and Agricultural Chemistry

	(iii) Seed Technology
	(iv) Agricultural Microbiology
	(v) Crop Physiology
	(vi) Agricultural Engineering
	(vii) Nano Science and Technology
	(viii) Agricultural Meteorology
	(ix) Veternary and Animal Sciences
	(x) Environmental Science
C. Crop Protection	(i) Agricultural Entomology
	(ii) Plant Pathology
	(iii) Nematology
D. Social Sciences	(i) Agricultural Extension
	(ii) Agricultural Economics
	(iii) Agricultural Resource Management
Allied disciplines	(i) Agricultural Statistics
	(ii) English and Tamil/Hindi/Telugu/Kannada/ Regional language
	(iii) Mathematics
	(iv) Bio-Chemistry

Evolution of Man and Agriculture

There are different stages in development of agriculture, which is oriented with human civilization. They are Hunting Pastoral Crop culture Trade (stages of human civilization).

1. **Hunting** - It was the primary source of food in old days. It is the important occupation and it existed for a very long period.

2. **Pastoral** - Human obtained his food through domestication animals, e.g. dogs, horse, cow, buffalo, etc. They lived in the periphery of the forest and they had to feed his domesticated animals. For feeding his animals, he would have migrated from one place to another in search of food. It was not comfortable and they might have enjoyed the benefit of staying in one place near the riverbed.

3. **Crop culture** - By living near the river bed, he had enough water for his animals and domesticated crops and started cultivation. Thus he has started to settle in a place.

4. **Trade** – When he started producing more than his requirement the excess was exchanged, this is the basis for trade. When agriculture

has flourished, trade developed. This lead to infrastructure development like road, routes, etc.

Agriculture became civilized from crop culture stage. Some important events for different periods that lead to development of scientific agriculture are :

Period	Events
Earlier than 10000 BC	Hunting & gathering
7500 BC	Cultivation of crops- Wheat & Barley
3400 BC	Wheel was invented
3000 BC	Bronze used for making tools
2900 BC	Plough was invented, irrigated farming started
2300 BC	Cultivation of chickpea, cotton, mustard
2200 BC	Cultivation of rice
1500 BC	Cultivation of sugarcane
1400 BC	Use of iron
1000 BC	Use of iron plough
1500 AD	Cultivation of orange, brinjal, pomegranate
1600 AD	Introduction of several crops in India i.e. potato, tapioca, tomato, chillies, pineapple, groundnut, tobacco, rubber, American cotton

Development of Scientific Agriculture in the World and India

Excavations, legends and remote sensing tests reveal that agriculture is 10,000 years old. Women by their intrinsic insight first observed that plants come up from seeds. Men concentrated on hunting and gathering (Paleolithic and Neolithic periods) during that time. Women were the pioneers for cultivating useful plants from the wild flora. They dug out edible roots and rhizomes and buried the small ones for subsequent harvests. They used animal meat as main food and their skin for clothing.

Shifting cultivation

A primitive form of agriculture in which people working with the crudest of tools, cut down a part of the forest, burnt the underneath growth and started new garden sites. After few years when these plots lost their fertility or choked with weeds or became heavily infested with soil-borne pests, they would shift to a new site. This system supported a spare population where abundant land is available. It is faulty land use.

This is also known as assartage system (cultivating crops till the land is completely worn-out) contrary to the fallow system. Fallow system means land is allowed for a resting period without any crop. Shifting cultivation practiced in Assam is referred as jhum cultivation, podu in Andhra Pradesh and Orissa, kumari in Western Ghats, watra in south east Rajasthan and panda bewar in Madhya Pradesh.

Subsidiary farming

Rudimentary system of settled farming since then agricultural activity became subsidiary to gathering and hunting. People in groups started settling down near a stream or river as permanent village sites and started cultivating in the same land more continuously; however the tools, crops and cropping methods were primitive.

Subsistence farming

Advanced form of primitive agriculture based on the principle of "Grow it and eat it" instead of growing crops on a commercial basis. Raising the crops only for family needs.

Mixed farming

It is the farming comprising of field crops and livestock. Field crop-grass husbandry (same field was used both for cropping and later grazing) was common. It is a stage changing from food gathering to food growing.

Advanced farming

Advanced farming practices included seed selection, green manuring with legumes, crop rotation, use of animal and crop refuse as manures, irrigation, pasture management, rearing of milch animals, bullocks, sheep and goat for wool and meat, rearing of birds by stall feeding etc.,

Modern agriculture

Started during 18th century with crop sequence, organic recycling, introduction of exotic crops and animals, use of farm implements in agriculture

Agriculture as Real Science / Scientific Agriculture: (19th Century)

- Research and development in fundamental and basic sciences were brought under applied aspects of agriculture.

- Agriculture took the shape of a teaching Science.
- Laboratories, farms, Research stations, Research centres, Institutes for research, teaching and extension (training and demonstration) were developed.
- Books, journals, popular and scientific articles, literatures were introduced.
- New media, and audio-visual aids were developed to disseminate new research findings and information to the rural masses.

Present day Agriculture (20th and 21st Century)

Today agriculture is not merely production oriented but is becoming a business consisting of various enterprises like livestock (dairy), poultry, fishery, piggery, sericulture, apiary, plantation cropping etc. Now a lot of developments on hydrological, mechanical, chemical, genetical and technological aspects of agriculture are in progress. Governments are apportioning a greater share of national budget for agricultural development. Small and marginal farmers are being supplied with agricultural inputs on subsidy. Policies for preserving, processing, pricing, marketing, distributing, consuming, exporting and importing are strengthening. Agro-based small scale industries and crafts are fast developing. Need based agricultural planning; programming and execution are in progress.

Agriculture – Development Scenario

Agriculture System	Cultural stage or Time	Average Cereal yield (t/ha)	World population (millions)	Per capital land availability (Hectares)
Hunting and Food Gathering	Paleolithic (old stone age about 2,50,000 to 10,000 BC)	-	7	-
Shifting Agriculture	Neolithic (new stone age about 7,000 B.C)	1	35	40.0
Medieval Agriculture	500-1450 A.D.	1	900	01.5
Livestock farming	18th Century	2	1800	00.7
Fertilizer / Pesticide Use in Agriculture	20th Century	4	4200	00.3

Development of Scientific Agriculture in World

Experimentation technique was started (1561 to 1624) by Francis Bacon. He conducted an experiment and found that water is the principle requirement for plant. If the same crop is cultivated for many times fertility is lost.

Jan Baptiste Van Helmont (1572-1644) was actually responsible for conducting a pot experiment. The experiment is called as 'willow tree experiment'. He took a willow tree of weight 5 pounds. He planted in a pot and the pot contained 200 pounds of soil and continuously monitored for five years by only watering the plant. By the end of 5th year, the willow tree was weighing 16 pounds. The weight of soil is 198 pounds. He concluded that water is the sole requirement for plants. The conclusion was erroneous.

In the 18th century, Arthur Young (1741-1820) published 'Annuals of Agriculture'.

Chronological Events in Scientific Agriculture

1561-1624 A.D : Francis Bacon	Found the water as nutrient of plants
1604-1668 A.D : G.R.Glanber	Salt peter (KNO_3) as nutrient and not water
1674-1741 A.D : Jethro Tull	Fine soil particle as plant nutrient, Perfected horse drawn seed drill (horse-hoeing husbandry)
1730-1799 A.D : Joseph Priestly	Discovered the oxygen
1775 A.D : Francis Home	Water, air, salts, fire and oil form the plant nutrients
1793 AD : Thomas Jefferson	Developed mould board plough
Theodore de-Saussure	Plants absorb CO_2 from air & release O_2; Soil supply N_2
1804- 1873 : Justus van Liebig	German chemist developed "law of minimum".

Liebig is a German scientist and considered as the 'Father of agricultural chemistry'. It was his opinion that the growth of plant was proportional to the amount of mineral substances available in the soil. This is called as 'Liebig law of minimum'. It states that, "A deficiency or absence of the necessary constituent, all others being present, renders the soil barren for crops for which that nutrient is needed" – It is referred as "Barrel concept".

If the barrel has stones of different heights, the lowest one establishes the capacity of the Barrel. Nitrogen has the lowest share, establishes the maximum capacity of the barrel. Accordingly, the growth factor in lowest supply (whether climatic, edaphic, genetic or biotic) sets the capacity for yield. Similarly a soil deficient in nitrogen (N) can't be made to produce well by adding more calcium (Ca) or potassium (K) where they are already abundant.

Advances in Agriculture in 19th Century

Following Liebig, an agricultural experiment station was started in Rothamsted in England on 1843 (Old Permanent Manurial Experiment – OPME), it dealt with nutrients. Subsequently many developments took place. In U.S. land grant colleges was started in 19th century. Its objective was to meet the expenditure of the college from the land around the colleges. USDA (United States Department of Agriculture) is responsible for the introduction of herbicides 2,4-D and tractor combine for harvesting and threshing. Under Land Grant College, agriculture oriented teaching, research, extension are expanded. Many international research institutes were started for a specific crop.

In 1875, Michigan State University was established to provide agriculture education on college level. Gregor Johann Mendel (1866) discovered the laws of heredity. Charles Darwin (1876) published the results of experiments on cross and self-fertilization in plants.

Thomas Malthus (1898) Proposed "Malthusian Theory" that the human race would run or later run out of food for everyone in spite of the rapid advances being made in agriculture at that time, because of limited land and yield potential of crops. Neo Malthusians have proposed birth control as answer to the problem.

F.T.Blackman's (1905) Theory of "Optima and Limiting Factors" states that, "when a process is conditioned as to its rapidity by a number of separate factors, the rate of the process is limited by the pace of the slowest factor'.

E.A. Mitsherlich (1909) proposed a theory of "Law of diminishing returns" states that 'the increase in any crop produce by a unit increment of a deficient factor is proportional to the decrement of that factor from the maximum and the response is curvilinear instead of linear'.

Mitscherlich equation is $dy/dx = C\ (A\text{-}Y)$ where,

d – Increment or change

dy – amount of increase in yield

dx – amount of increment of the growth factor x.
A – Maximum possible yield
Y – Yield obtained for the given quantity of factor 'x' and
C – Proportionality constant that depends on the nature of the growth factor.

Wilcox (1929) proposed "Inverse Yield – Nitrogen law" states that, the growth are yielding ability of any crop plant is inversely proportional to the mean nitrogen content in the dry matter.

Macy (1936): Proposed a concept of "Critical Percentages of Plant Nutrients" He suggested a relationship between the sufficiency of nutrients and plant response in terms of yield and nutrient concentration of plant tissues". Macy proposed critical percentages for each nutrient in each kind of plant.In the tissues minimum percentage range, an added increment of a nutrient increases the yield but not the nutrient percentage. In the poverty adjustment range, an added increment of a nutrient increases the nutrient percentage but not the yield. In the luxury consumption range, added increment of nutrient have little effect of yield. But increase the nutrient composition percentage. The point between poverty adjustment and luxury consumption was the "Critical percentage"

Macy suggested that Liebig's law holds well in the tissue minimum percentage range because there is not enough of a nutrient to allow much plant growth. Liebig's law holds good again in the luxury consumption range. Because there is a large supply of nutrient, some other nutrient becomes limiting and stops growth. Mitscherlich's law of diminishing returns holds during the poverty adjustment range because the response curve is linear representing the diminishing yield to added increments.

Zimmerman and Hitchcock (1942) reported that 2,4-D could act as growth promoter at extremely low concentration. Now 2,4-D is used to overcome the problem of seediness in Poovan banana.

In 1945, herbicide 2,4,5-T was developed. In 1954, Gibberllic acid structure was identified by Japanese. In 1950's Bennet and Clark identified ABA (Abscissic acid) which inhibits plant growth and controls shedding of plant parts.

Development of Scientific Agriculture in India

Scientific agriculture got momentum in the 19th century itself. Indian Land Tax was levied in the middle of 19th century. In 1877, 1878, 1889, 1892, 1897 and 1900, the population was decreased due to continuous famines. Only due to these famines, the British regime started many development programmes. Lord Dalhousie (1848-1856) period the 'Upper

Bari Doab Canal' in Punjab was constructed. Improvement of agriculture started only in his period. In Lord Curzon's (1898-1905) period, the 'Great Canal system of Western Punjab' was constructed. During his period Imperial Agricultural Research Institute was started in Pusa in Bihar. His period is called as 'Golden period of agriculture'. During his regime, Department of Agriculture and Agricultural colleges for provinces were started at Coimbatore in 1906.

Due to earthquake at IARI in Pusa, Bihar it was shifted to New Delhi. In 1926, Royal Commission on Agriculture was setup and was responsible for giving recommendation to dug canals, lay roads, etc. Based upon the recommendation of Royal Commission, ICAR (Imperial Council of Agricultural Research) was started in 1929 with the objective to conduct agriculture research. State Agricultural Universities (SAU were started after 1960s). ICAR had also started research institutes of its own in different centres in India for various crops.

ICAR is the sole body, which controls all the Agricultural Research Institutes in India. It paved way for green revolution in India. After 1947, ICAR totally adapted to Land Grant Colleges. In 1962, a Land Grant College was started in Pantnagar (UP). It is the first university with 16,000 acres.

There are 45 state agricultural universities with research institutes on their own. High yielding wheat varieties like Kalyansona, Sonalika, Lerma roja and Sonara-64 were introduced. Green revolution took in wheat first, next in rice after the invention of Indo-*Japanica* variety. Today, agricultural research is multi-dimensional. It includes tissue culture, biotechnology besides breeding, crop production and crop protection.

Milestones

1880 - Department of Agriculture was established

1903 - Imperial Agricultural Research Institute (IARI) was started at Pusa, Bihar

1912 - Sugarcane Breeding Institute was established in Coimbatore

1929 - Imperial Council of Agricultural Research at New Delhi (then ICAR) after independence becomes ICAR

1936 - Due to earth quake in Bihar, IARI was shifted to New Delhi and the place was called with original name Pusa

1962 - First Agricultural University was started at Pantnagar

1965-67 Green revolution in India due to introduction of HYV -Wheat, rice, use of fertilizers, construction of Dams and use of pesticides

In Tamilnadu

1876 - Madras Agricultural College was established at Saidapet

1906 - Agricultural College & Research Institute was established at Coimbatore

1971 - Tamil Nadu Agricultural University was started.

For institutes visit http://www.icar.org.in/node/325

For state agricultural universities visit http://www.icar.org.in/en/universities.htm

National Institutions for Agricultural Research

1. ICAR - National Rice Research Institute, Cuttack
2. ICAR - Vivekananda Parvatiya Krishi Anusandhan Sansthan, Almora
3. ICAR - Indian Institute of Pulses Research, Kanpur
4. ICAR - Central Tobacco Research Institute, Rajahmundry
5. ICAR - Indian Institute of Sugarcane Research, Lucknow
6. ICAR - Sugarcane Breeding Institute, Coimbatore
7. ICAR - Central Institute of Cotton Research, Nagpur
8. ICAR - Central Research Institute for Jute and Allied Fibres, Barrackpore
9. ICAR - Indian Grassland and Fodder Research Institute, Jhansi
10. ICAR - Indian Institute of Horticultural Research, Bangalore
11. ICAR - Central Institute of Sub Tropical Horticulture, Lucknow
12. ICAR - Central Institute of Temperate Horticulture, Srinagar
13. ICAR - Central Institute of Arid Horticulture, Bikaner
14. ICAR - Indian Institute of Vegetable Research, Varanasi
15. ICAR - Central Potato Research Institute, Shimla
16. ICAR - Central Tuber Crops Research Institute, Trivandrum
17. ICAR - Central Plantation Crops Research Institute, Kasargod
18. ICAR - Central Agricultural Research Institute, Port Blair
19. ICAR - Indian Institute of Spices Research, Calicut
20. ICAR - Central Soil and Water Conservation Research & Training Institute, Dehradun
21. ICAR - Indian Institute of Soil Sciences, Bhopal

22. ICAR - Central Soil Salinity Research Institute, Karnal
23. ICAR - ICAR Research Complex for Eastern Region including Centre of Makhana, Patna
24. ICAR - Central Research Institute of Dryland Agriculture, Hyderabad
25. ICAR - Central Arid Zone Research Institute, Jodhpur
26. ICAR - Central Coastal Agriculture Research Institute, Goa
27. ICAR - ICAR Research Complex for NEH Region, Barapani
28. ICAR - National Institute of Abiotic Stress Management, Malegaon Maharashtra
29. ICAR - Central Institute of Agricultural Engineering, Bhopal
30. ICAR - Central Institute on Post harvest Engineering and Technology Ludhiana
31. ICAR - Indian Institute of Natural Resins and Gums, Ranchi
32. ICAR - National Institute of Research on Jute & Allied Fibre Technology, Kolkata
33. ICAR - Indian Agricultural Statistical Research Institute, New Delhi
34. ICAR - Indian Institute of Farming System Research, Modipuram, ICAR - Meerut
35. ICAR - Central Sheep and Wool Research Institute, Avikanagar, Rajasthan
36. ICAR - Central Institute for Research on Goats, Makhdoom
37. ICAR - Central Institute for Research on Buffaloes, Hissar
38. ICAR - National Institute of Animal Nutrition and Physiology, Bangalore
39. ICAR - Central Avian Research Institute, Izatnagar
40. ICAR - Central Marine Fisheries Research Institute, Kochi
41. ICAR - Central Institute of Brackishwater Aquaculture, Chennai
42. ICAR - Central Inland Fisheries Research Institute, Barrackpore
43. ICAR - Central Institute of Fisheries Technology, Cochin
44. ICAR - Central Institute of Freshwater Aquaculture, Bhuvaneshwar
45. ICAR - National Academy of Agricultural Research & Management, Hyderabad

National Research Centres - 7

1. ICAR - National Research Centre on Plant Biotechnology, New Delhi
2. ICAR - National Centre for Integrated Pest Management, New Delhi
3. ICAR - National Research Centre for Litchi, Muzaffarpur
4. ICAR - National Research Centre for Citrus, Nagpur
5. ICAR - National Research Centre for Grapes, Pune
6. ICAR - National Research Centre for Banana, Trichirappalli, Tamil Nadu
7. ICAR - National Research Centre for Seed Spices, Ajmer
8. ICAR - National Research Centre for Pomegranate, Solapur
9. ICAR - National Research Centre on Orchids, Pakyong, Sikkim
10. ICAR - National Research Centre for Agroforestry, Jhansi
11. ICAR - National Research Centre on Camel, Bikaner
12. ICAR - National Research Centre on Equines, Hissar
13. ICAR - National Research Centre on Meat, Hyderabad
14. ICAR - National Research Centre on Pig, Guwahati
15. ICAR - National Research Centre on Yak, West Kemang
16. ICAR - National Research Centre on Mithun, Medziphema, Nagaland
17. ICAR - National Centre for Agril. Economics & Policy Research, New Delhi
18. ICAR - National Organic Farming Research Institute, Sikkim.

Important International Institutions on Agricultural Research

AVRDC - Asian Vegetable Research and Development Centre, Taiwan

CIAT - Centro International de Agricultura Tropical, Cali, Colombia

CIP - Centro International da la Papa (International Potato Research Institute (Lima, Peru, South America)

CIMMYT - Centro International de Mejoramiento de Maizy Trigo. (International Centre for Maize and Wheat Development (Londress, Mexico)

IITA - International Institute for Tropical Agriculture, Ibadon in Nigeria, Africa)

ICARDA - International Centre for Agricultural Research in the Dry Areas (Aleppo, Syria)

ICRISAT - International Crops Research Institute for the Semi Arid Tropics (Pattancheru in Hyderabad, India)

IIMI - International Irrigation Management Institute (Colombo, Srilanka)

IRRI - International Rice Research Institute (Los Banos, Phillippines)

ISNAR - International Service in National Agricultural Research (The Hague, Netherlands)

WARDA - West African Rice Development Association (Ivory Coast, Africa)

IBPGR - International Board for Plant Genetic Resources (Rome, Italy)

IFPRI - International Food Policy Research Institute

CIFOR - Center for International Forestry Research

IWMI - International Water Management Institute

ICRAF - World Agroforestry Centre

CGIAR - Consultative Group on International Agricultural Research (Washington D.C)

FAO - Food and Agricultural Organization (Rome)

WMO - World Meteorological Organization (Vienna)

Indian Agriculture and Economy

Indian Agriculture is one of the most significant contributors to the Indian economy. Agriculture is the only means of living for almost 60% of the employed class in India. The agriculture sector of India has occupied almost 43% of India's geographical area. Agriculture is still the only largest contributor to India's GDP (16%) even after a decline in the same in the agriculture share of India. Agriculture also plays a significant role in the growth of socio-economic sector in India.

In the earlier times, India was largely dependent upon food imports, but the successive story of the agriculture sector of Indian economy has made it self-sufficing in grain production. The country also has substantial reserves for the same. India depends heavily on the agriculture sector, especially on the food production unit after the 1960 crisis in food sector. Since then, India has put a lot of effort to be self-sufficient in the food production and this endeavour of India has led to the Green Revolution.

The Green Revolution came into existence with the aim to improve the agriculture in India.

The services enhanced by the Green Revolution in the agriculture sector of Indian economy are as follows:

- Acquiring more area for cultivation purposes
- Expanding irrigation facilities
- Use of improved and advanced high-yielding variety of seeds
- Implementing better techniques that emerged from agriculture research
- Water management
- Plan protection activities through prudent use of fertilizers, pesticides.

All these measures taken by the Green Revolution led to an alarming rise in the wheat and rice production of India's agriculture. Considering the quantum leap witnessed by the wheat and rice production unit of India's agriculture, a National Pulse Development Programme that covered almost 13 states was set up in 1986 with the aim to introduce the improved technologies to the farmers. A Technology Mission on Oilseeds was introduced in 1986 right after the success of National Pulse Development Programme to boost the oilseeds sector in Indian economy. Pulses too came under this programme. A new seed policy was planned to provide entry to superior quality seeds and plant material for fruits, vegetables, oilseeds, pulses and flowers.

The Indian government also set up Ministry of Food Processing Industries to stimulate the agriculture sector of Indian economy and make it more lucrative. India's agriculture sector highly depends upon the monsoon season as heavy rainfall during the time leads to a rich harvest. But, the entire year's agriculture cannot possibly depend upon only one season. Taking into account this fact, a second Green Revolution is likely to be formed to overcome such restrictions. An increase in the growth rate and irrigation area, improved water management, improving the soil quality and diversifying into high value outputs, fruits, vegetables, herbs, flowers, medicinal plants and bio-diesel are also on the list of the services to be taken by the Green Revolution to improve the agriculture in India.

National income

National Income is important because of the following reasons :

- To see the economic development of the country.
- To assess the developmental objectives.
- To know the contribution of various sectors to national income.

Internationally some countries are wealthy, some countries are not wealthy and some countries are in-between. Under such circumstances, it would be difficult to evaluate the performance of an economy. Performance of an economy is directly proportionate to the amount of goods and services produced in an economy. Measuring national income is also important to chalk out the future course of the economy. It also broadly indicates people's standard of living. Income can be measured by Gross National Product (GNP), Gross Domestic Product (GDP), Gross National Income (GNI), Net National Product (NNP) and Net National Income (NNI).The Indian economy is the 12th largest in USD exchange rate terms. India is the second fastest growing economy in the world. India's GDP has touched US$1.25 trillion. The crossing of Indian GDP over a trillion dollar mark in 2007 puts India in the elite group of 12 countries with trillion dollar economy. The tremendous growth rate has coincided with better macroeconomic stability. India has made remarkable progress in information technology, high end services and knowledge process services.

Agricultural income in GDP

Agriculture sector contributed 32% in 1990-91, 20% during 2005-06 and around 16% now. Though the contribution of agriculture to the GDP income of India, it is great news that today the service sector is contributing more than half of the Indian GDP. It takes India one step closer to the developed economies of the world. Earlier it was agriculture which mainly contributed to the Indian GDP. The Indian government is still looking up to improve the GDP of the country and so several steps have been taken to boost the economy. Policies of FDI, SEZs and NRI investment have been framed to give a push to the economy and hence the GDP.

Agriculture per capita income

The per capita income of the agriculture sector declines to 1/3 of the national per capita income during the recent years. The per capita income of the agriculture population is estimated around Rs. 10,865 in 2010, which is around 32% of the national per capita income (Rs. 33,802/-).

Income Distribution in Agriculture Sector

Year/ Period	Agriculture Share in GDP (%)	Population Dependent on Agriculture (Millions)	Agriculture Per Capita (Rs.)
1980	39	70	4745
1990	31	65	5505
2000	25	59	6652
2010	16	58	10865

Women in Agriculture and Empowerment

Women in India now participate in all activities such as education, sports, politics, media, art and culture, service sectors, science and technology, etc. Indira Gandhi, who served as Prime Minister of India for an aggregate period of fifteen years is the world's longest serving woman Prime Minister.

The Constitution of India guarantees to all Indian women equality (Article 14), no discrimination by the State [Article 15(1)], equality of oppurtunity (Article 16) and equal pay for equal work [Article 39(d)]. In addition, it allows special provisions to be made by the State in favour of women and children [Article 15(3)], renounces practices derogatory to the dignity of women [Article 51(a) (e)], and also allows for provisions to be made by the State for securing just and humane conditions of work and for maternity relief (Article 42).

The feminist activism in India picked up momentum during later 1970s. Since alcoholism is often associated with violence against women in India, many women groups launched anti-liquor campaigns in Andhra Pradesh, Himachal Pradesh, Haryana, Orissa, Madhya Pradesh and other states.

In 1990s, grants from foreign donor agencies enabled the formation of new women-oriented NGOs. Self-help groups and NGOs such as Self Employed Women's Association (SEWA) have played a major role in women's rights in India. Many women have emerged as leaders of local movements. For example, Medha Patkar of the Narmada Bachao Andolan.

The Government of India declared 2001 as the Year of Women's Empowerment (*Swashakti*). The National Policy for the Empowerment of Women came was passed in 2001.

In 2010 March 9, one day after International Women's day, Rajyasabha passed Women's Reservation Bill, ensuring 33% reservation to women in Parliament and state legislative bodies.

Women empowerment would become more relevant if women are educated, better informed and can take rational decisions. It is also necessary to sensitize the other sex towards women. It is important to usher in changes in societal attitudes and perceptions with regard to the role of women in different spheres of life. Adjustments have to be made in traditional gender specific performance of tasks. A woman needs to be physically healthy so that she is able to take challenges of equality. But it is sadly lacking in a majority of women especially in the rural areas. They have unequal access to basic health resources.

Most of the women work in agricultural sector either as workers, in household farms or as wageworkers. Yet it is precisely livelihood in agriculture that has tended to become more volatile and insecure in recent years and women cultivators have therefore been negatively affected. The government's policies for alleviating poverty have failed to produce any desirable results, as women do not receive appropriate wages for their labour. There is also significant amount of unpaid or non-marketed labor within the household. The increase in gender disparity in wages in the urban areas is also quite marked as it results from the employment of women in different and lower paying activities. They are exploited at various levels. They should be provided with proper wages and work at par with men so that their status can be elevated in society.

There is no doubt about the fact that development of women has always been the central focus of planning since Independence. Empowerment is a major step in this direction but it has to be seen in a relational context. A clear vision is needed to remove the obstacles to the path of women's emancipation both from the government and women themselves. Efforts should be directed towards all round development of each and every section of Indian women by giving them their due share.

Questions

I. Fill in the blanks

1. The term Agriculture is derived from Latin words ȃger or agri means ______ and cultura means ______.
2. In India, contribution to the gross domestic product (GDP) through agriculture is ____%.
3. Law of minimum concept was proposed by ______.
4. ________ period is called as Golden period of agriculture
5. Vedic period in India is during ________.

II. Choose the correct answer

6. Imperial Agricultural Research Institute at Pusa (Bihar) was started during
   a. 1881 b. 1903
   c. 1929 d. 1962
7. Imperial Council of Agricultural Research (ICAR) was established during
   a. 1881 b. 1905
   c. 1929 d. 1962
8. Green revolution in India due to introduction of HYV was during
   a. 1960-61 b. 1965-66
   c. 1970-71 d. 1980-81
9. National Rice Research Institute is situated at
   a. Cuttack b. Hyderabad
   c. Coimbatore d. New Delhi
10. Earliest tools of farmers are made of
   a. Wood b. Stone
   c. Both a & b d. None of these

III. Write short answer

1. Agriculture is a science
2. Subsistence farming
3. Inverse Yield - Nitrogen law
4. Green Revolution
5. Pastoral stage.

2

Agricultural Heritage in India

History is the continuous record of past events

Heritage is the inherited values carried from one generation to other generation

Agricultural heritage refers to the values and traditional practices adopted in ancient India which are more relevant for present day system.

Agricultural Heritage in India

Agriculture in India is not of recent origin, but has a long history dating back to Neolithic age of 7500-6500 B.C. It changed the life style of early man from 'nomadic hunter of wild berries and roots' to 'cultivator of land'. Agriculture is benefited from the wisdom and teachings of great saints. The wisdom gained and practices adopted have been passed down through generations. The traditional farmers have developed the nature friendly farming systems and practices such as mixed farming, mixed cropping, crop rotation etc. The great epics of ancient India convey the depth of knowledge possessed by the older generations of the farmers of India. The modern society has lost sight of the importance of the traditional knowledge which had been subjected to a process of refinement through generations of experience. The ecological considerations shown by the traditional farmers in their farming activities are now-a-days is reflected in the resurgence of organic agriculture.

The available ancient literature includes the four Vedas (rig, yajur, sama, atharvana), nineteen Brahmanas (A total of 19 Brahmanas are extent at least in their entirety: two associated with the Rigveda, six with the Yajurveda, ten with the Samaveda and one with the Atharvaveda.), Aranyakas, Sutra literature, Susruta Samhita, Charaka Samhita, Upanishads, the epices Ramayana and Mahabharata, Puranas (20), Buddhist and Jain literature, and texts such as Krishi-Parashara, Kautilya's Arthasastra, Panini's Ashtadhyayi, Sangam literature of Tamils,

Manusmirti (laws), Varahamihira's Brihat Samhita (maths & astrology), Amarkosha, Kashyapiya-Krishisukti and Surapala's Vriskshayurveda. This literature was most likely to have been composed between 6000 BC and 1000 AD. The information related to the biodiversity and agriculture (including animal husbandry) is available in these texts.

Rigveda is the most ancient literary work of India. It believed that Gods were the foremost among agriculturists. According to Amarakosha (a thesaurus of Sanskrit written by the Jain or Buddhist scholar Amarasimha), Aryans were agriculturists. Manu and Kautilya prescribed agriculture, cattle rearing and commerce as essential subjects, which the king must learn. According to Patanjali (compiler of the Yoga Sûtras) the economy of the country depended on agriculture and cattle-breeding. Plenty of information is available in 'Puranas', which reveals that ancient Indians had intimate knowledge on all agricultural operations. Some of the well known ancient classics of India are namely, Kautilya's' Arthashastra'; Panini's 'Astadhyayi'; Patanjali's 'Mahabhasya'; Varahamihira's 'Brahat Samhita'; Amarsimha's 'Amarkosha' and Encyclopaedic works of Manasollasa. These classics testify the knowledge and wisdom of the people of ancient period. Technical books dealing exclusively with agriculture were Sage Parashara's 'Krishiparashara' in 1000 A.D. Other important texts are Agni Purana and Krishi Sukti attributed to Kashyap (500 A.D.). Ancient Tamil and Kannada works contain lot of useful information on agriculture in ancient India. Agriculture in India made tremendous progress in the rearing of sheep and goats, cows and buffaloes, trees and shrubs, spices and condiments, food and non-food crops, fruits and vegetables and developed nature friendly farming practices. These practices had social and religious undertones and became the way of life for the people. Domestic rites and festivals often synchronised with the four main agricultural operations of ploughing, sowing, reaping and harvesting.

In the Rigveda, there is reference to hundreds and thousands of cows; to horses yoked to chariots; to race courses where chariot races were held; to camels yoked to the chariots; to sheep and goats offered as sacrificial victims, and to the use of wool for clothing. The famous Cow Sukta indicates that the cow had already become the very basis of rural economy. In another Sukta, she is defined as the mother of the Vasus, the Rudras and the Adityas, as also the pivot of Immortality. The Vedic Aryans appear to have large forests at their disposal for securing timber, and plants and herbs for medicinal purposes appear to have been reared by the physicians of the age, as appears in the Atharva Veda. The farmers' vocation was held in high regard, though agriculture solely depended upon the favours of Parjanya, the god of rain. His thunders are described as food-bringing.

The four Vedas mention more than 75 plant species, Satapatha Brahamna mentions over 25 species, and Charkaa Samhita (300 BC) an Ayurvedic (Indian medicine) treatise-mentions more than 320 plants. Susruta (400 BC) records over 750 medicinal plant species. The oldest book, Rigveda (4000 BC) mentions a large number of poisonous and non poisonous aquatic and terrestrial, and domestic and wild creatures and animals. Puranas mention about 500 species of plants.

The science of arbori-horticulture had developed well and has been documented in Surapala's Vrikshayurveda. Forests were very important in ancient times. From the age of Vedas, protection of forests was emphasized for ecological balance. Kautilya in his Artha Sastra (321-296 BC) mentions that superintendent of forests had to collect forest produce through the forest guards. He provides a long list of trees, varieties, of bamboos, creepers, fibrous plants, drugs and poisons, skins of various animals, etc., that come under the purview of this officer.

The preservation of wild animals was encouraged and hunting as a sport was regarded as detrimental to proper development of the character and personality of the ruler, according to Manu (Manusmriti, 2nd Century BC). Specifically, in the Puranas (300-750 AD) the names of Shalihotra on horses and Palakapya on elephants have been found as experts in animal husbandry. For instance, Garudapurana is a text dealing with treatment of animal disorders while the classical work on the treatment of horses is Aswashastra. One chapter in Agnipurana deals with the treatment of livestock and another on treatment of trees.

Stages of Agriculture Development

12000 to 9500 years ago

- Hunters and food-gatherers stage existed.
- Stone implements (microliths) were seen throughout the Indian subcontinent.
- Domestication of dog occurred in Iraq.
- Earliest agriculture was by vegetative propagation (e.g., banana, sugarcane, yam, sago, palms, and ginger).

9500 to 7500 years ago

- Wild ancestors of wheat and barley, goat, sheep, pig, and cattle were found.

7500 to 5000 years ago

- Significant features were invention of plough, irrigated farming, use of wheel, and metallurgy and in Egypt, seed dibbling said to be practiced.

5000to 4000 years ago

- Harappan culture is characterized by cultivation of wheat, barley and cotton; plough agriculture and bullocks for drought. Indus Valley is the home of cotton.
- Wheeled carts were commonly used in the Indus valley.
- Harappans not only grew cotton but also devised methods for ginning / spinning / weaving.

4000 to 2000 years ago

- In North Arcot, bone / stone tools were found.
- In Nevasa (Maharashtra), copper and polished stone axes were used. First evidence of the presence of silk was found at this location.
- At Navdatoli on Narmada river (Nemar, Madhya Pradesh), sickles set with stone teeth were used for cutting crop stalks. Crops grown were wheat, linseed, lentil, urd (black gram), mung bean, and khesari.
- In Eastern India, rice, bananas, and sugarcane were cultivated.

2000-1500 years ago

- Tank irrigation was developed and practiced widely.
- Greek and Romans had trade with South India; pepper, cloth, and sandal wood were imported by Romans.
- Chola King Karikala (190 AD) defeated Cheras and Pandyas, invaded Srilanka, captured 12000 men and used them as slaves to construct an embankment along the Cauvery, 160 km along, to protect land from floods. He has built numerous irrigation tanks and promoted agriculture by clearing forests.

1500-1000 years ago

The Kanauj Empire of Harshavardhana (606-647 AD)

- Cereals such as wheat, rice and millets, and fruits were extensively grown. A 60-day variety and fragrant varieties of rice are mentioned.

- Ginger, mustard, melons, pumpkin, onion, and garlic are also mentioned.
- Persian wheel was used in Thanesar (Haryana).

The Kingdoms of South India

- The kingdoms were of the Chalukyas (Badami), Rashtrakutas (Latur), Pallavas (Kanchi), Pandyas, Hoysals (Helebid), and Kakatiyas (Warangal).
- Cholas ushered in a glorious phase in South Indian in the 10th century AD.
- New irrigation systems for agriculture were developed- chain tanks in Andhra in the 9th century; and 6.4 km Kaveripak bund.
- Cholas maintained links with China, Myanmar, and Campodia.
- The tank supervision committee (Eri-variyam) looked after the maintenance of a village and regulated the water supply.

1000-700 years ago

- Arab conquest of Sind was during 711-712 AD; Md bin Qasim defeated Dahir, the Hindu king of Sind. Arabs were experts in gardening.
- 1290- 1320AD (Reign of Khiljis): Alauddin Khilji destroyed the agricultural prosperity of a major part of India. He believed in keeping the farmers poor.

Era of civilization

It is supposed that man was evolved on earth about 1.7 million years ago. This man was evolved from the monkey who started to move by standing erect on his feet. Such man has been called *Homo erectus* (or) Java man (or) Peking man. Later on Java man transformed into Cro-Magnon and Cro-Magnon into modern man. The modern man is zoologically known as *Homo sapiens* (Homo - Continuous, Sapiens - learning habit). A primitive form of *Homo sapiens,* called Neanderthal man (*Homo sapiens neanderthalensis*), was common in Europe and Asia. After the last glacial period (about 10,000 years ago), modern *Homo sapien sapiens,* began to spread all over the globe.

In the beginning such man had been spending his life wildly, but during the period 8700-7700 BC, they started to pet sheep and goat, although the first pet animal was dog, which was used for hunting. The history of agriculture and civilization go hand in hand as the food production made it possible for primitive man to settle down in selected areas leading to formation of society and initiation of civilization. The development of civilization and agriculture had passed through several stages. Archaeologist initially classified the stages as stone age, Bronze and Iron age. Subsequently the scholars spilt up the stone age into Paleolithic period (old stone age), Neolithic age (New stone age) and Mesolithic age (Middle stone age).

Each of three ages, saw distinct improvements. The man fashioned and improved tools out of stones, bones, woods etc. to help them in day-to-day life. They started growing food crops and domesticated animals like cow, sheep, goat, dog etc.

I. The Stone Age culture: (2500 BC to 3500 BC)

The stone age is divided into three periods

A. Palaeolithic period (old stone age)

B. Mesolithic period (middle stone age)

C. Neolithic period (new stone age)

A. **Paleolithic age:** Hunters and food gatherers (2,50,000 to 10,000 BC)

This period is characterized by the food gatherers and hunters. The stoneage man started making stone tools and crude choppers. The chipped stone tools and chopped pebbles were used for hunting, cutting and other purposes. He had no knowledge on cultivation and house building.

The Palaeolithic age in India is divided into three phases according to the nature of stone tools used by the people and according the nature of climate

a) Early or lower palaeolithic (2,50,000 to 1,00,000 BC)

b) Middle palaeolithic (1,00,000 to 40,000 BC)

c) Upper palaeolithic (40,000 to 10,000 BC)

B. **Mesolithic period:** Hunters and Herders (10,000 to 3,700 BC)

The transitional period between the end of the Paleolithic and beginning of the Neolithic is called Mesolithic. It began about 10000BC and ended with the rise of agriculture. This period is characterized by tiny stone implements called microliths. The

Mesolithic people lived on hunting, fishing and good-gathering. At later stages, they also domesticated animals. The domestication of the dog was the major achievement of the Mesolithic hunter.

C. **Neolithic Age:** Food producers (The beginning of Agriculture)

The Neolithic age began between 9000 and 7500 BC. Neolithic revolution occurred in Western Asia between 9500 and 8500 years ago mainly in the **Fertile Crescent** (hilly regions embracing Israel, Jordan, Turkey, Iran, Caspian basin and adjoining Iranian plateau). Neolithic revolution brought a major change in the techniques of food production which gave man control over his environment and saved him from the precarious existence of mere hunting and gathering of wild berries and roots. For the first time, he lived in **settled villages** and apart from security from hunger he had leisure time to think and contemplate.

The main features of Neolithic culture in India

1. Neolithic culture denotes a stage in economic and technological development in India
2. Use of polished stone axes for cleaning the bushes
3. Hand made pottery for storing food grains
4. Invented textile, weaving and basketry
5. Cultivation of rice, banana sequence and yams in eastern parts of India
6. Cultivation of millets and pulses in south India
7. Discovery of silk.

II. Bronze Age (Chalcolithic culture) : (3000-1700 BC)

The end of the Neolithic period saw the use of metal. The metal to be used first was copper. The term Chalcolithic (stone – copper phase) is applied to communities using stone implements along with copper and bronze. In more advanced communities, the proportion of copper and bronze implements is higher than that of stones. The chalcolithic revolution began in Mesopotamia in the fourth millennium B.C. from this area it spread to Egypt, and Indus valley.

The significant features are

1. Invention of plough
2. Agriculture shifted from hilly area to lower river valley
3. Flood water were stored for irrigation and canals were dug
4. Irrigated farming started in this period
5. Sowing of seed by dibbling with a pointed stick
6. Use of wheel.
7. Salinity problem and water logging were noticed due to canal irrigation.

III. The Iron Age

The Iron phase in India started after the advent of Aryans. Aryan land was called *Saptasindhava* i.e. the land of seven rivers (Sutlej, Beas, Ravi, Chenab, Jhelum, Sindhu and Saraswati). Aryans settle along the rivers.

The salient features of iron age are

- The harvesting tool used was **sickle.**
- The crops grown were mainly sesame and sugarcane.
- Iron plough, shear, axe, sickles and hoes were used
- The domestic animals were sheep, goats, dogs, mule, tortoise, cattle breeds and elephants.

Chronology of Agricultural Technology Development in India

Agriculture in India is broadly classified in to five different periods before India's independence.

1. Before 1500 BCE : Early history
2. 1500 BCE - 200 CE : Vedic period – Post Maha Janapadas period
3. 200–1200 CE : Early Common Era – High Middle Ages
4. 1200–1757 CE : Late Middle Ages – Early Modern Era
5. 1757–1947 CE : Colonial British Era

*[**Note:** BCE - short for "Before the Common Era", "Before the Christian Era", or "Before the Current Era". CE - Common Era, Current Era (Christian Era is, however, also abbreviated AD, for Anno Domini]*

Indian agriculture began by 9000 BCE as a result of early cultivation of plants and domestication of crops and animals. Settled life soon followed with implements and techniques being developed for agriculture. Double monsoons led to two harvests being reaped in one year. Indian products soon reached the world via existing trading networks and foreign crops were introduced to India. Plants and animals; considered essential to their survival by the Indians, came to be worshipped and venerated.

The middle ages saw irrigation channels reach a new level of sophistication in India and Indian crops affecting the economies of other regions of the world under Islamic patronage. Land and water management systems were developed with an aim of providing uniform growth. Despite some stagnation during the later modern era the independent Republic of India was able to develop a comprehensive agricultural program.

1. Early history (Before 1500 BCE)

- **9000 BCE:** Wheat and barley were domesticated in the Indian subcontinent. Domestication of horse, sheep and goat soon followed. This period also saw the first domestication of the elephant.
- **8000-6000 BCE:** Barley and wheat cultivation, along with the domestication of cattle, primarily sheep and goat—was visible in Mehrgarh (Baluchistan, now in Pakistan). Agro pastoralism in India included threshing, planting crops in rows—either of two or of six—and storing grain in granaries.
- **5000 BCE:** Agricultural communities became widespread in Kashmir.
- **5000-4000 BCE:** Cotton was cultivated. The Indus cotton industry was well developed and some methods used in cotton spinning and fabrication continued to be practiced till the modern Industrialization of India. A variety of tropical fruit such as mango and muskmelon are native to the Indian sub-continent. The Indians also domesticated hemp, which they used for a number of applications including making narcotics, fibre and oil. The farmers of the Indus Valley grew peas, sesame, and dates. Sugarcane was originally from tropical South Asia and Southeast Asia. Different species likely originated in different locations with *S. barberi* originating in India and *S. edule* and *S. officinarum* coming from New Guinea.

- **5440 BCE:** Wild *Oryza* rice appeared in the Belan and Ganges valley regions of northern India. Rice was cultivated in the Indus Valley civilization.
- **4500 BCE:** Irrigation was developed in the Indus Valley Civilization. The size and prosperity of the Indus civilization grew as a result of this innovation, which eventually led to more planned settlements making use of drainage.
- **3000 BCE**: Sophisticated irrigation and water storage systems were developed by the Indus Valley civilization, including artificial reservoirs at Girnar.
- **2600 BCE:** An early canal irrigation system from Circa.
- **2500 BCE:** Archeological evidence of an animal-drawn plough in the Indus Valley civilization
- **2000 BCE:** Agricultural activity included rice cultivation in the Kashmir and Harrappan regions.

2. Vedic period-Post Maha Janapadas period (1500 BCE–200 CE)

- Gupta (2004) finds it likely that summer monsoons may have been longer and may have contained moisture in excess than required for normal food production. One effect of this excessive moisture would have been to aid the winter monsoon rainfall required for winter crops.
- In India, both wheat and barley are held to be *Rabi* (winter) crops and—like other parts of the world—would have largely depended on winter monsoons before the irrigation became widespread. The growth of the *Kharif* crops would have probably suffered as a result of excessive moisture.
- Jute was first cultivated in India, where it was used to make ropes and cordage.
- Some animals—thought by the Indians as being vital to their survival—came to be worshipped.
- Trees were also domesticated, worshipped, and venerated—*Pipal* and *Banyan* in particular.
- Others came to be known for their medicinal uses and found mention in the holistic medical system *Ayurveda*.
- **1000–500 BCE:** There are repeated references to iron. Cultivation of a wide range of cereals, vegetables and fruits is described. Meat

and milk products were part of the diet; animal husbandry was important. The soil was ploughed several times. Seeds were broadcasted. Fallowing and a certain sequence of cropping were recommended. Cow dung provided the manure. Irrigation was practiced.

- **322–185 BCE:** The Mauryan Empire categorized soils and made meteorological observations for agricultural use. Other Mauryan facilitation included construction and maintenance of dams, and provision of horse-drawn chariots—quicker than traditional bullock carts.
- **300 BCE:** The Greek diplomat Megasthenes, in his book *Indica*—provides a secular eyewitness account of Indian agriculture.

3. Early Common Era – High middle ages (200–1200 CE)

- The Tamil people cultivated a wide range of crops such as rice, sugarcane, millets, black pepper, various grains, coconut, beans, cotton, plantain, tamarind and sandalwood. Jackfruit, coconut, palm, areca and plantain trees were also known.
- Systematic ploughing, manuring, weeding, irrigation and crop protection was practiced for sustained agriculture. Water storage systems were designed during this period.
- Kallanai (1st-2nd century CE), a stone dam built on river Kaveri during this period, is considered as one of the oldest water regulation structures in the world still in use.
- Spice trade involving spices native to India—including cinnamon and black pepper—gained momentum as India starts shipping spices to the Mediterranean.
- Roman trade with India followed as detailed by the archaeological record and the *Periplus of the Erythrean Sea.*
- Chinese sericulture attracted Indian sailors during the early centuries of the Common Era.
- **320-550 CE:** Crystallized sugar was discovered by the time of the Guptas and the earliest reference of candied sugar come from India.
- **647 CE:** Chinese documents confirm at least two missions to India, initiated in, for obtaining technology for sugar-refining.
- **875-1279 CE:** Noboru Karashima's research of the agrarian society in South India during the Chola Empire reveals that during the

Chola rule land was transferred and collective holding of land by a group of people slowly gave way to individual plots of land, each with their own irrigation system.

- The growth of individual disposition of farming property may have led to a decrease in areas of dry cultivation.
- The Cholas also had bureaucrats which oversaw the distribution of water-particularly the distribution of water by tank-and-channel networks to the drier areas.

4. Late middle ages – Early modern era (1200–1757 CE)

- The construction of water works and aspects of water technology in India is described in Arabic and Persian works. The diffusion of Indian and Persian irrigation technologies gave rise to irrigation systems which bought about economic growth and growth of material culture.
- Agricultural 'zones' were broadly divided into those producing rice, wheat or millets.
- Rice production continued to dominate Gujarat and wheat dominated north and central India.
- The Encyclopædia Britannica details the many crops introduced to India during this period of extensive global discourse.
- **1556-1605 CE:** Land management was particularly strong during the regime of Akbar the Great under whom scholar-bureaucrat Todarmal formulated and implemented elaborated methods for agricultural management on a rational basis.
- Indian crops—such as cotton, sugar, and citric fruits—spread visibly throughout North Africa, Islamic Spain, and the Middle East.
- Though they may have been in cultivation prior to the solidification of Islam in India, their production was further improved as a result of this recent wave, which led to far-reaching economic outcomes for the regions involved.

5. Colonial British Era (1757–1947 CE)

- A number of irrigation canals are located on the Sutlej river.
- Few Indian commercial crops—such as Cotton, indigo, opium, and rice—made it to the global market under the British Raj in India.

- The second half of the 19th century saw some increase in land under cultivation and agricultural production expanded at an average rate of about 1% per year by the later 19th century.
- Due to extensive irrigation by canal networks Punjab, Narmada valley, and Andhra Pradesh became centers of agrarian reforms.
- The British regime in India did supply the irrigation works but rarely on the scale required.
- Community effort and private investment soared as market for irrigation developed.
- Agricultural prices of some commodities rose to about three times between 1870-1920.
- A rich source of the state of Indian agriculture in the early British era is a report prepared by a British engineer, Thomas Barnard, and his Indian guide, Raja Chengalvaraya Mudaliar, around 1774. This report contains data of agricultural production in about 800 villages in the area around Chennai in the years 1762 to 1766. This report is available in Tamil in the form of palm leaf manuscripts at Thanjavur Tamil University, and in English in the Tamil Nadu State Archives.
- 1871: British Empire created Department of Revenue, Agriculture and Commerce, which formed as base for initiation of Agriculture in India.
- 1880: Famine Commission Report was submitted which was the base for inception of Agricultural Department.
- 1881: Separate Department of Agriculture at Centre for Famine relief operations
- 1890: Dr. J.A. Voelcker appointed as a consulting chemist from Royal Agricultural Society (England) - Laid foundation for agricultural research in India.
- 1892 – 1903 - Appointment of Imperial Agricultural Chemist, Imperial Mycologist and Imperial Entomologist – Base for Beginning of inducting the scientist in Agriculture.
- 1901-05: To enhance agricultural education, Establishment of Agricultural Colleges at Pune, Kanpur, Sabour, Nagpur, Coimbatore and Lyallpur (Now in Pakistan).
- 1905: Establishment of Imperial Agricultural Research Institute (IARI) at Pusa (Bihar)

- 1929: Based on Royal Commission on Agriculture's recommendation (1928), Imperial Council of Agricultural Research (ICAR) was establishment to conduct comprehensive research.
- 1931-47: Indian Lac Cess Committee, Indian Central Tobacco Committee, Indian Central Oilseeds Committee were formed to improve research in various crops.

Republic of India (1947 CE onwards)

- Special programs were undertaken to improve food and cash crops supply.
- The Grow More Food Campaign (1940s) and the Integrated Production Programme (1950s) focused on food and cash crops supply respectively.
- 1957: All India Coordinated Maize Improvement Project was initiated (First coordinated project) to exploit maize research (Specifically heterosis).
- Five-year plans of India - oriented towards agricultural development—soon followed.
- 1963: Introduction of semi dwarf wheat varieties from CIMMYT, Mexico Formed basis for green revolution.
- 1966: Introduced semi-dwarf rice varieties TN1 & IR 8 from Taiwan and Philippines respectively is formed as base for green revolution.
- Land reclamation, land development, mechanization, electrification, use of chemicals—fertilizers in particular, and development of agriculture oriented 'package approach' of taking a set of actions instead of promoting single aspect soon followed under government supervision.
- The many 'production revolutions' initiated from 1960s onwards included Green Revolution in India, Yellow Revolution (oilseed: 1986-1990), Operation Flood (dairy: 1970-1996), and Blue Revolution (fishing: 1973-2002) etc.
- 1979: National Agricultural Research Project (NARP) was launched to strengthen the research capabilities of SAUs
- Following the economic reforms of 1991, significant growth was registered in the agricultural sector, which was by now benefiting from the earlier reforms and the newer innovations of Agro-processing and Biotechnology.

- 1998: National Agricultural Technology Project (NATP) was initiated Strengthen the research on location specific problems Contract farming – which requires the farmers to produce crops for a company under contract – and high value agricultural product increased.
- 2006: National Agricultural Innovative Project (NAIP) was launched for end to end approach for solving problems

Agriculture in Arthasasthra

Kautilya (also known as Vishnu Gupta or Chanakya) (321-296 BC) was a great scholar of time. He wrote a treatise titled, Arthasasthra, which deals with the management of resources. During Kautilya's time agriculture, cattle breeding and trade were grouped into a science called 'Varta'. Kautilya gave great importance to agriculture and suggested a separate post of head of agriculture and named as 'Sitadhakashya'. Agriculture today receives prime importance, by policy and administrative support from government officials. eg. i) Supply of good seeds and other inputs, ii) Provision of irrigation water, iii) prediction of rainfall by IMD, iv) Assistance in purchase of machineries, v) Marketing and safe storage. All the important aspects are mentioned by Kautilya in his book. He suggested many important aspects in agriculture which are highly relevant today.

1. The superintendent of agriculture should be a person who is knowledgeable in agriculture and horticulture. There was a provision to appoint a person who was not an expert but he was assisted by other knowledgeable person.
2. Anticipation of labours by land owners before sowing. Slaves and prisoners were organised to sow the seeds in time. He also emphasized that ploughing provides good soil texture required for a particular crop.
3. Timely sowing is very important for high yield particularly for rainfed sowing for which, all the implements and accessories have to be kept ready. Any delay in these arrangements received punitive action.
4. Kautilya suggested that for getting good yield of rainfed crop, a rainfall of 16 dronas (One drona = 40 mm to 50 mm; so totally 600-800 mm) was essential and 40 dronas rainfall (1600-200 mm) is sufficient for rice. It is very significant to note that rain gauge was used during Kautilya's period. It was apparently a circular vessel

(20 fingers width, 8 fingers depth) and the unit to measure rain was adhaka (1 adhaka=12 mm approx.)

5. He also stressed the optimum distribution of rainfall during crop growing season. One third of the required quantity of rainfall must fall both in the commencement (July/August) and closing months (October-December) of rainy season; and 2/3 of rainfall in the middle (August-October) is considered as very even.

6. The crops should be sown according to the the season. E.g. *Sali* (transplant rice), *Virlu* (direct sown rice), *Till* (Sesame) and millets should be sown at the commencement of rain. Pulses to be sown in the middle of season. Safflower, linseed, mustard, barley and wheat to be sown later.

7. He also stressed that rice crop require less labour expense, vegetables are intermediate and sugarcane is worst as it requires more attention and expenditure.

8. The crops like cucurbits are well suited to banks of rivers, Long-peper, sugarcane and grapes do well where the soil profile is well charged with water. Vegetable require frequent irrigation, borders of field suited for cultivation of medicinal plants.

9. Some of the bio-control practices suggested by Kautilya have got relevance.They are:

   a) Practice of exposing seeds to mist and heat for seven nights. These practices are followed even now in wheat to prevent smut diseases.

   b) Cut ends of sugarcane are plastered with the mixture of honey, ghee and cow-dung. Recently evidences proved that honey has wide antimicrobial property. Ghee could seal off the cut ends prevent loss of moisture and cow-dung facilitated bio-control of potential pathogens.

10. He also suggested that harvesting should be done at proper time and nothing should be left in the field not even chaff. The harvested produce should be properly processed and safely stored. The above ground crop residues were also removed from fields and fed to cattle.

Agriculture in the Sangam Literature

During the Sangam period (200 BC to 100 AD), the main profession of the population of the Tamil region (now Tamil Nadu) was agriculture.

The region extended from Cape Comorin (Kaniyakumari) in the South to Tirupati (in Andhra Pradesh) in the North, parts of present Kerala and Karnataka in the West. The methods of cultivation practised during this ancient period were revealed by several proverbs, village songs and literature of the period which are available even today. It is rather surprising that the people had good knowledge about agriculture (seed varieties, seed selection, seed storage, ploughing, manuring, irrigation, weeding, crop protection, pests, and botanical pesticides).

The Sangam period literature covers wide aspects of the people's life, such as epics, ethics, social life and religion. Several poems composed during this period have been passed on from generation to generation through memorizing and chanting and later through manuscripts written on *Palmyara* leaves. With the advent of paper and printing machinary, Shri Swaminatha Iyer who is popularly called 'Tamill grandfather' painstakingly collected them and brought them out as printed books. Two peoms of the Sangam period, viz., Tholkappiyam and Thirukural, gives us a vivid picture of agricultural practices in that period.

Tholkappiyam

The poem Tholkappiyam was written by the poet Tholkappier during 200 BC. It gives descriptions of various agricultural aspects and are enumerated below.

Land classification

Land was classified into five (but, cultivable land in to four) groups, viz., *Mullai* (forest), *Kurinji* (hills), *Marudham* (cultivable lands), and *Neithal* (coastal areas). *Palai* land was not brought under cultivation and left as fallow.

Seasons

Six seasons are mentioned: Early spring, late spring, cloudy, rainy, early winter, and late winter.

Cultivated crops

There are references to rice, millets, sugarcane, banana, cardamom, pepper, cotton, sesame, coconut and nut. Farmers were aware that rice could be grown as rainfed crops. Banana and sugarcane were ratooned. Plants were considered as living beings and endowed with sensitivity. Tholkappier also mentioned about monocots and dicots.

Importance of agriculture

Kings considered agricultural development as their primary duty. They felt that soil fertility and irrigation facilities should be the country's assets. Increased agricultural production was considered a yardstick of prosperity of the country. The stability of a kingdom was ensured not by army but by agriculture and sufficient crop production. Failure of monsoon rains and reduction in grain yield were attributed to the king's sins.

Irrigation

Kings dug-out tanks at locations where water flow from rains was plentiful. Semicircular bunds were raised adjacent to small hillocks and water reservoirs a kin to present day dams were raised and constructed. It indicates awareness of water harvesting. The king 'Karikal Cholan' brought 1000 slaves from a conquered country and raised the bunds of river Cauvery. The stone dam constructed across the river Cauvery centuries ago is considered a master piece of engineering even today. River water was diverted to tanks through canals. It is mentioned that irrigation should be given both in early morning or late evening, and not during hot mid-day.

Agricultural implements

Buffaloes were used for ploughing with a wooden plough. Deep ploughing was considered superior to shallow ploughing. A labour saving tool called *Parambu* was used for leveling paddy fields. Tools such as *Amiry, Keilar,* and *Yettam* were used to lift water from wells, tanks, and rivers. Tools called *Thattai* and *Kavan* were used for scaring birds in millet fields. Traps were used to catch wild boars in millet fields.

Seeds

Seed was selected from those earheads that first matured. The selected seed was stored for sowing only and never used as food grain. It was believed that such a diversion would destroy the family.

Crop rotation

Crop rotation was practised by raising black gram (urd) after rice. This indicates that farmers were aware of the benefits to the following rice crop which we now know is due to the nitrogen fixation in the root nodules of urd. They also practised mixed cropping; e.g., foxtail millet with lablab or cotton.

Threshing

A tool called *Senyam* was used for harvesting rice. Threshing of rice was done by hand with the help of a buffalo (and in large holdings by elephants). Hand winnowing was done to remove chaff. One sixth of the produce was paid as tax to the king. Farm labourers were paid in kind.

The land was immediately ploughed after harvest or water was allowed to the field to facilitate rooting of stubbles. Operations requiring hard work such as ploughing were done by men while women attended to light work such as transplanting, weeding, bird scaring, harvesting and winnowing. In Kandapuranam, it is mentioned that Valli, daughter of a king, was sent for bird scaring in millet fields where Lord Muruga (son of Lord Shiva) courted her and married.

Marketing

Products were exchanged by weight. In Madurai (the headquarters of Sangam poets), there was a food grain bazaar where 18 kinds of cereals, millets and pulses were sold. Each shop had a banner hoisted high so that it could be seen from a distance indicating that the grains are sold here. Customs duty was collected on imports and exports.

Thirukural

The poem was composed by a gifted poet named *Thiruvalluvar* during 70 BC. It consists of 1330 couplets (133 topics each having 10 couplets). It is the pride of Sangam Tamil literature and its greatness can be realized from the fact that it has been translated into English and several other languages. It devotes one topic (10 couplets) for agriculture under the chapter politics. This clearly reveals the recognition that the prime duty of a king is to ensure agricultural production.

Importance of agriculture

'World spins around many industries. All such industries spin around agriculture'

'Farmers alone live an independent life; others worship them and are second to them'. 'If farmers stop cultivation, even rishis (sages) cannot survive'

Ploughing

'If land is ploughed deep and soil allowed drying to one fourth weight, even manuring is not necessary'

Manuring

'Manuring is more important than ploughing: crop protection is more important than irrigation'. Green leaf manuring, farmyard manure, and sheep penning were in vogue though farmers were not aware that they supplied nitrogen to the crop. One is amazed at the depth of agricultural knowledge our ancestors possessed.

Irrigation

Bed method was followed as an efficient method of water management.

Weeding

'Just like the farmer pulls out weeds with the root system, so the king should eliminate criminals from society'.

Care of crops

'If the farmer does not regularly visit his field, the crop will not grow'

The foregoing account of agriculture from ancient Tamil literature clearly indicates the agricultural knowledge of our forefathers. By following their footsteps, the present generation of agricultural scientists has used the advanced technologies and has tried to stabilize agricultural production in our country to meet our food requirements.

Rainfall Prediction

- Large number of fireflies seen at night on the forest trees is a sign that the monsoon will start early (Farmers in Maharashtra).
- If there is rain, accompanied with lightning and mild thunder on the second day of Jayastha month (May – June), there will be no rain for the next 72 days (Farmers Gujarat)

Indigenous Technical Knowledge (ITK)

ITK is defined as the sum total of knowledge and practices which are based on people's accumulated experience in dealing with situations and problems in various aspects of life and such knowledge and practices are special to a particular culture.

When the farmers continuously practicing indigenous knowledge, it will be also relevant to enquire why they do so. In other words, what are the advantages of such practices as perceived by farmers? Understanding the rational of such practices from farmers' point of view, may also help researchers to look into the valid factors while they research to farmers need and help extension workers to select appropriate technologies based on few criteria

- Summer ploughing conserves moisture, eradicates weeds, consolidates soil erosion and minimizes the number of ploughings at the time of sowing.
- Due to cowdung coating for cotton seeds, the easy dibbling of seeds to remove fuzz, good germination, no cost and pest-reduction were the advantages.
- Soaking sorghum seeds in cow urine before sowing increase the drought tolerance and the seeds had germination with minimum rain and it was considered as no cost practice.
- Soaking Bengal gram in water as found with the previous practice, farmers had resorted to the practice of soaking bengal gram in water before sowing because it considered as no cost and withstanding water stress.
- Cotton seeds treated with red soils facilitate easy dibbling of seeds and it favours good germination.
- Cattle penning practices improve the soil fertility owing to organic manure.
- Sorghum mixed with lab-lab given additional yield owing to mixed cropping and it enhances nitrogen fixation by leguminous lab-lab.
- Use of cow dung cake as burrow fumigant is economical in controlling rats.
- Raising castor as a border crop in cotton field is used as trap crop for the cotton pest and it also provided additional income.
- Easy removal of pest, easy separation of kernels, longer shelf life and higher economics are advantages of coating red gram with red soil.
- Mixing green gram with ash was one of the post harvest indigenous practice and it had certain advantages like pest reduction and cheaper method.

Tamil Almanac (*Panchangam*)

An annual publication including weather forecasts and other miscellaneous information arranged according to the calendar of a given year. The Tamil Almanac is used in Tamil Nadu and Puducherry in India, and by the Tamil population in Malaysia, Singapore and Sri Lanka. It is used today for cultural, religious and agricultural events, with the Gregorian calendar having largely supplanted it for official use both within and outside India. It is based on the classical Hindu solar calendar also used in Assam, Bengal, Kerala, Manipur, Nepal, Orissa and the Punjab.

There are several festivals based on the Tamil Hindu calendar. The Tamil New Year follows the *Nirayanam* vernal equinox and generally falls on April 13 or 14th of the Gregorian year. April 13 or 14th marks the first day of the traditional Tamil calendar and this remains a public holiday in both Tamil Nadu and Sri Lanka. Tropical vernal equinox fall around 22 March, and adding 23 degrees of trepidation or oscillation to it, we get the Hindu sidereal or *Nirayana Mesha Sankranti* (Sun's transition into *nirayana* Aries). Hence, the Tamil calendar begins on the same date in April which is observed by most traditional calendars of the rest of India.

Week

The days of the Tamil calendar relate to the celestial bodies in the solar system: Sun, Moon, Mars, Mercury, Jupiter, Venus, and Saturn, in that order. The week starts with Sunday.

This list compiles the days of the week in the Tamil calendar:

No.	Weekday (Tamil)	Sanskrit name	Lord or Planet	Gregorian Calendar equivalent
01.	*Gnyaayitru-kizhamai*	Ravivaara	Sun	Sunday
02.	*Thingat-kizhamai*	Somavaara	Moon	Monday
03.	*Sevvaai-kizhamai*	Mangalavaara	Mars	Tuesday
04.	*Buthan-kizhamai*	Budhavaara	Mercury	Wednesday
05.	*Viyaazha-kizhamai*	Guruvaara	Jupiter	Thursday
06.	*Velli-kizhamai*	Sukravaara	Venus	Friday
07.	*Sani-kizhamai*	Shanivaara	Saturn	Saturday

Months

The number of days in a month varies between 29 and 32.

The following list compiles the months of the Tamil Calendar

No.	Month (Tamil)	Sanskrit Name	Gregorian Calendar equivalent
01.	*Cittirai*	*Chaitra*	mid-April to mid-May
02.	*Vaikaci*	*Vaisakha*	mid-May to mid-June
03.	*Aani*	*Jyaishtha*	mid-June to mid-July
04.	*Aadi*	*Ashadha*	mid-July to mid-August
05.	*Aavani*	*Shravana*	mid-August to mid-September
06.	*Purattashi*	*Bhadrapada*	mid-September to mid-October
07.	*Aippaci / Aippasi*	*Ashwina*	mid-October to mid-November
08.	*Karttikai*	*Karttika*	mid-November to mid-December
09.	*Marka*	*Margashirsha*	mid-December to mid-January
10.	*Tai*	*Pausha*	mid-January to mid-February
11.	*Maci*	*Magha*	mid-February to mid-March
12.	*Pankuni*	*Phalguna*	mid-March to mid-April

Seasons

The Tamil year, in keeping with the old Indic calendar, is divided into six seasons, each of which lasts two months

Season name	English translation	Sanskrit Name	English equivalent	Months
Kar	Dark, rain	Varsha	Rainy	Aavani, Purataci
Kutir	Chill, wind	Sharada	Autumn	Aippaci, Kârthikai
Munpani	Early dew	Hemanta	Early winter	Markazhi, Tai
Pinpani	Late dew	Sishira	Late winter	Masi, Pankuni
Ilavenil	Young warmth	Vasanta	Spring	Chithirai, Vaikasi
Mutuvenil	Extreme warmth	Grishma	Summer	Aani, Aadi

Questions

I. Fill in the blanks

1. Sage Parashara's technical book dealing exclusively with agriculture is ______
2. The poem ____________________ classified the land into five groups

3. The oldest water regulation structure in the world is ____________.
4. Puranas mention about ____________ species of plants.
5. Neolithic revolution occurred in Western Asia between 9500 and 8500 years ago mainly in the ____________.

II. Choose the correct answer

6. Old stone age is called as

   a. Palaeolithic period b. Mesolithic period

   c. Neolithic period d. None

7. Large number of poisonous and non-poisonous plants are mentioned in

   a. Charkaa Samhita b. Rig veda

   c. Susruta d. Puranas

8. As per Kautilya, the quantity of rainfall essential for getting good yield of rainfed crop is

   a. 12 dronas b. 16 dronas

   c. 20 dronas d. 24 dronas

9. The chalcolithic revolution in the fourth millennium B.C. began in

   a. Mesopotamia b. `Mohanjadaro

   c. Harappa d. Nile

10. Text dealing treatment of animal disorders during Kautilya period is

    a. Garudapurana b. Aswashastra

    c. Agnipurana d. Shalihotra

III. Answer the following

1. Agricultural Heritage
2. Rainfall in Arthasasthra
3. Bio-control practices suggested by Kautilya
4. Kallanai
5. Rainfall prediction.

□□□

3

Agronomy and Agroclimatic Zones

The word agronomy has been derived from two Greek words, **agros** and **nomos** having the meaning of **field** and to **manage**, respectively. Literally, agronomy means the "art of managing field". Technically, it means the "Science and economics of crop production by management of farm land".

Definitions

Agronomy is the art and underlying science in production and improvement of field crops with the efficient use of soil fertility, water, labour and other factors related to crop production.

Agronomy is defined as "a branch of agricultural science which deals with principles and practices of field crop production and management of soil for higher productivity.

Scope of Agronomy

Agronomy is a dynamic discipline with the advancement of knowledge and better understanding of planet, environment and agriculture. Agronomy science becomes imperative in Agriculture in the following areas.

- Identification of proper season for cultivation of wide range of crops is needed which could be made possible only by Agronomy science.
- Proper methods of cultivation are needed to reduce the cost of cultivation and maximize the yield and economic returns.
- Availability and application of chemical fertilizers has necessitated the generation of knowledge to reduce the ill-effects due to excess application and yield losses due to the unscientific manner of application.

- Availability of herbicides for control of weeds has led to development for a vast knowledge about selectivity, time & method of its application.
- Water management practices play greater role in present day crisis of water demand and Agronomy science answer to the questions 'how much to apply?' and 'when to apply?'.
- Intensive cropping is the need of the day and proper time and space intensification not only increase the production but also reduces the environmental hazards.
- New technology to overcome the effect of moisture stress under dry land condition is explored by Agronomy and future agriculture depends on dry land agriculture.
- Packages of practices to explore full potential of new varieties of crops are the most important aspects in crop production which could be made possible only by Agronomy science.
- Keeping farm implements in good shape and utilizing efficient manner to nullify the present day labour crisis is further broadening the scope of agronomy.
- Maintaining the ecological balance through efficient management of crops, livestock and their feedings in a rational manner is possible only by knowing agronomic principles.
- Care and disposal of farm and animal products like milk and eggs and proper maintenance of accounts of all transactions concerning farm business is governing principles of agronomy.

Dimensions of Agronomy

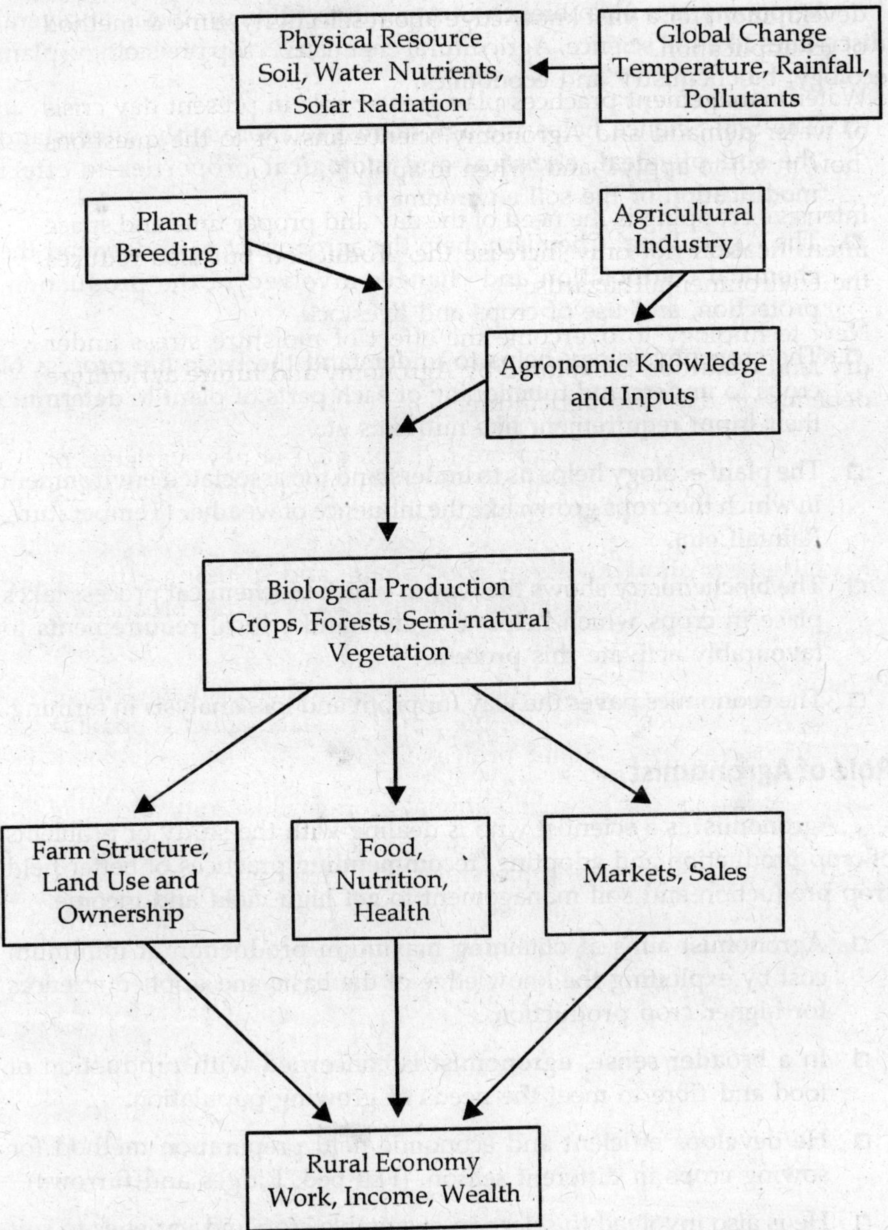

Flow Diagram of Physical, Biological, Economic and Social Dimensions of Agronomy.

Relation of Agronomy to Other Sciences

Agronomy is a main branch of Agriculture. It is synthesis of several disciplines like soil science, Agricultural chemistry, crop physiology, plant ecology, biochemistry and economics.

- The Soil Science helps the agronomist to thoroughly understand the soil physical, chemical and biological properties to effect modification of the soil environment.
- The Agricultural Chemistry help the agronomist to understand the chemical composition and changes involved in the production, protection, and use of crops and livestock.
- The crop physiology helps to understand the basic life process of crops to understand functioning of each parts of plant to determine their input requirement like nutrients etc.
- The plant ecology helps us to understand the associated environment in which the crops grown like the influence of weather (Temperature, Rainfall etc).
- The biochemistry shows the way in which biochemical process takes place in crops which helps to understand critical requirements to favourably activate this process.
- The economics paves the way for profit and loss analysis in farming.

Role of Agronomist

Agronomist is a scientist who is dealing with the study of problems of crop production and adopting/recommending practices of better field crop production and soil management to get high yield and income.

- Agronomist aims at obtaining maximum production at minimum cost by exploiting the knowledge of the basic and applied sciences for higher crop production.
- In a broader sense, agronomist is concerned with production of food and fibre to meet the needs of growing population.
- He develops efficient and economic field preparation method for sowing crops in different season. (Flat bed, Ridges and furrows)
- He is also involved to selection of suitable crop and varieties to suit or to match varied seasons and soils. Eg. Red soil - groundnut, Black soil - cotton, Sandy soil – tuberous crops, Saline soil – Finger millet (*Ragi*). In *Kharif* if water is sufficient go for rice and water is not sufficient go for maize, sorghum.

- Evolves efficient method of cultivation (weather broadcasting, nursery and transplantation or dibbling, etc.) provides better crop establishment and maintain required population
- He has to identify various types of nutrients required by crops including time and method of application (e.g. for long duration rice (150-60-60 kg NPK/ha), short duration: 120:50:50 kg NPK/ha Application P&K basal and N in three splits)
- Agronomist must select a better weed management practice. Either through mechanical or physical (by human work) or chemical (herbicides or weedicides, e.g. 2-4-D) or cultural (by having wide space it may increase weed growth by using inter space crops). Weeds are controlled by integrated weed management method also
- Selection of proper irrigation method, irrigation scheduling i.e. irrigation timing and quantity based on the crops to be irrigated, whether to irrigate continuously or stop in between and how much water to be supplied are computed by agronomy science so as to achieve maximum water use efficiency.
- Crop planning (i.e.) suitable crop sequence are developed by agronomist (i.e.) what type of crop, cropping pattern, cropping sequence, etc. (Rice - Rice - Pulse)
- Agronomists are also develops the method of harvesting, time for harvesting, etc. (Appropriate time of harvest essential to prevent yield loss)
- Agronomist is responsible for every decision made in the farm management. (What type of crop to be produced? How much area to be allotted for each crop? How and when to market? How and When to take other management activities?) All the decisions should be taken at appropriate time to efficiently use resources available).

Agro-climatic Zones

An agro-climatic zone is a land unit uniform in respect of climate and length of growing period (LGP) which is climatically suitable for a certain range of crops and cultivars (FAO, 1983).

Classification by Planning Commission

Planning Commission of India (1989) made an attempt to delineate the country into different agro climatic regions based on homogeneity in rainfall, temperature, topography, cropping and farming systems and water resources. India is divided into 15 agro-climatic regions.

1. Western Himalayan zone

This zone consists of three distinct sub-zones of Jammu and Kashmir, Himachal Pradesh and Uttar Pradesh hills. The region consists of skeletal soils of cold region, podsolic mountain meadow soils and hilly brown soils. Lands of the region have steep slopes in undulating terrain. Soils are generally silty loams and these are prone to erosion hazards.

2. Eastern Himalayan zone

Sikkim and Darjeeling hills, Arunachal Pradesh, Meghalaya, Nagaland, Manipur, Tripura, Mizoram, Assam and Jalpaiguri and Coch Bihar districts of West Bengal fall under this region, with high rainfall and high forest cover. Shifting cultivation is practiced in nearly one-third of the cultivated area and this has caused denudation and degradation of soils with the resultant heavy runoff, massive soil erosion and floods in lower reaches and basins.

3. Lower Gangetic Plains zone

This zone consists of West Bengal-lower Gangetic plain region. The soils are mostly alluvial and are prone to floods.

4. Middle Gangetic Plains zone

This zone consists of 12 districts of eastern Uttar Pradesh and 27 districts of Bihar plains. This zone has a geographical area of 16 million hectares and rainfall is high. About 39% of gross cropped area is irrigated and the cropping intensity is 142%.

5. Upper Gangetic Plains zone

This zone consists of 32 districts of Uttar Pradesh. Irrigation is through canals and tube wells. A good potential for exploitation of ground water exists.

6. Trans-Gangetic Plains zone

This zone consists of Punjab, Haryana, Union territories of Delhi and Chandigarh and Sriganganagar district of Rajasthan. The major characteristics of this area are: highest net sown area, highest irrigated area, high cropping intensity and high groundwater utilization.

7. Eastern Plateau and Hills zone

This zone consists of eastern part of Madhya Pradesh, southern part of West Bengal and most of inland Orissa. The soils are shallow and medium in depth and the topography is undulating with a slope of 1-10%. Irrigation is through tanks and tube wells.

8. Central Plateau and Hills zone

This zone comprises of 46 district of Madhya Pradesh, part of Uttar Pradesh and Rajasthan. The topography is highly variable nearly 1/3rd of the land is not available for cultivation. Irrigation and cropping intensity are low. 75% of the area is rainfed grown with low value cereal crops. There is an intensive need for alternate high value crops including horticultural crops.

9. Western Plateau and Hills zone

This zone comprises the major part of Maharashtra, parts of Madhya Pradesh and one district of Rajasthan. The average rainfall of the zone is 904 mm. The net sown area is 65% and forests occupy 11%. The irrigated area is only 12.4% with canals being the main source.

10. Southern Plateau and Hills zone

This zone comprises 35 districts of Andhra Pradesh, Karnataka and Tamil Nadu which are typically semi-arid zones. Dryland farming is adopted in 81% of the area and the cropping intensity is 111 percent.

11. East Coast Plains and Hills zone

This zone comprises of east coast of Tamil Nadu, Andhra Pradesh and Orissa. Soils are mainly alluvial and coastal sands. Irrigation is through canals and tanks.

12. West Coast Plains and Ghats zone

This zone comprises west coast of Tamil Nadu, Kerala, Karnataka, Maharashtra and Goa with a variety of crop patterns, rainfall and soil types.

13. Gujarat Plains and Hills zone

This zone consists of 19 districts of Gujarat. This zone is arid with low rainfall in most parts and only 32.5% of the area is irrigated largely through wells and tube wells.

14. Western Dry zone

This zone comprises nine districts of Rajasthan and is characterized by hot sandy desert, erratic rainfall, high evaporation, scanty vegetation. The ground water is deep and often brackish. Famine and drought are common features of the region.

15. Islands zone

This zone covers the island territories of Andaman and Nicobar and Lakshadeep which are typically equatorial with rainfall of 3000 mm spread over eight to nine months. It is largely a forest zone with undulated lands.

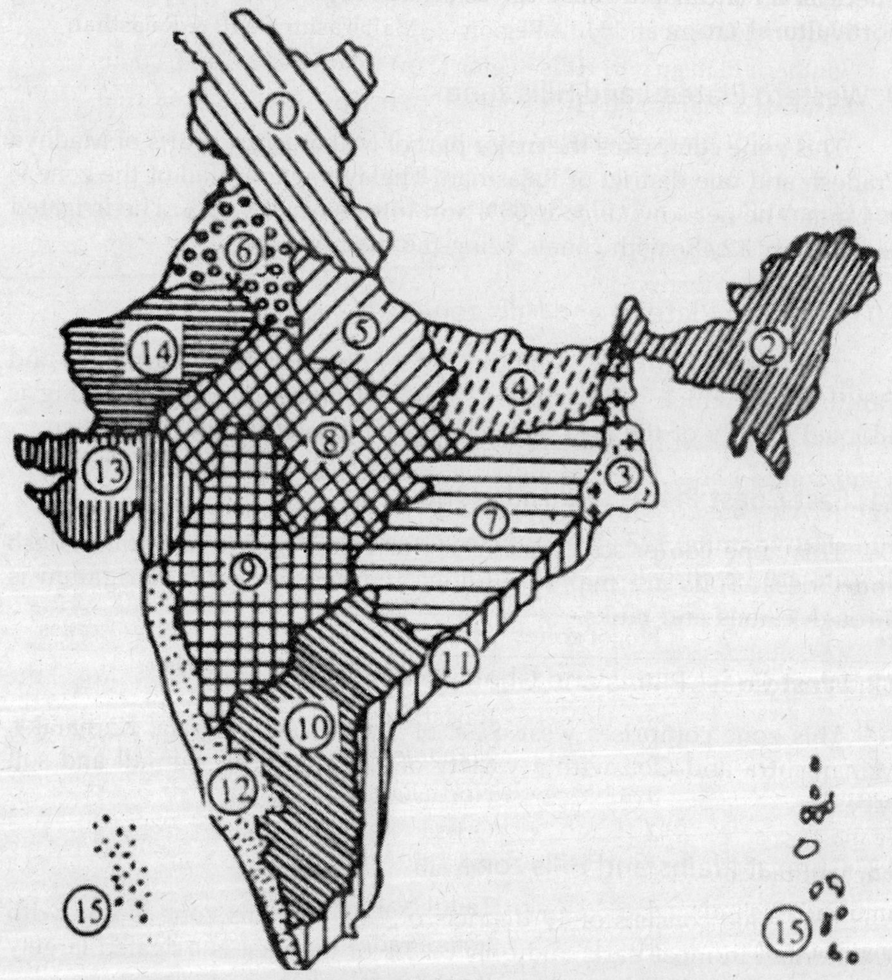

S.No.	Agroclimatic Zones	State
1	Western Himalayan Region	J&K, HP, UP, Uttarakhand
2	Eastern Himalayan Region	Assam Sikkim, West Bengal & North-Eastern states
3	Lower Gangetic Plains Region	West Bengal
4	Middle Gangetic Plains Region	UP, Bihar
5	Upper Gangetic Plains Region	UP
6	Trans-Gangetic Plains Region	Punjab, Haryana, Delhi & Rajasthan
7	Eastern Plateau and Hills Region	Maharashtra, UP, Orissa & West Bengal
8	Central Plateau and Hills Region	MP, Rajasthan, UP
9	Western Plateau and Hills Region	Maharashtra, MP & Rajasthan
10	Southern Plateau and Hills Region	AP, Karnataka, Tamil Nadu
11	East Coast Plains and Hills Region	Orissa, AP, TN,& Pondichery
12	West Coast Plains and Ghat Region	TN, Kerala, Goa, Karnataka, Maharashtra
13	Gujarat Plains and Hills Region	Gujarat
14	Western Dry Region	Rajasthan
15	The Islands Region	Andman & Nicobar, Lakshwadweep

Classification by ICAR

The State Agricultural Universities were advised to divide each state into sub-zones, under the National Agricultural Research Project (NARP) under ICAR. Based on the rainfall pattern, cropping pattern and administrative units, 127 agro-climatic zones are classified. The zones of each state are given below.

State	No. of zones	State	No. of zones
Andhra Pradesh	7	Madhya Pradesh	12
Assam	6	Rajasthan	9
Bihar	6	Maharashtra	9
Gujarat	8	North Eastern Hill region	6
Haryana	2	Orissa	9
Himachal Pradesh	4	Punjab	5
Jammu and Kashmir	4	Tamil Nadu	7
Karnataka	10	Uttar Pradesh	10
Kerala	8	West Bengal	6

Agroclimatic Zones of Tamilnadu

The state of Tamil Nadu has been classified into seven distinct agro-climatic zones listed below :

1. North Eastern zone
2. North Western zone
3. Western zone
4. Cauvery Delta zone
5. Southern zone
6. High Rainfall zone
7. High altitude and Hilly zone

1. North Eastern zone

This zone covers the districts of Thiruvallur, Vellore, Kanchipuram, Thiruvannamalai, Viluppuram, Cuddalore (excluding Chidambaram and Kattumannarkoil taluks), some parts of Perambalur including Ariyalur taluks and also Chennai. The mean annual rainfall of this region is 1054 mm received in 53 rainy days and is benefited by both the monsoons. The mean monthly maximum temperature ranges between 28.2 to 38.9 °C and the minimum ranges from 19.5 to 24.8°C.

2. North Western zone

This zone comprises of Dharmapuri and Krishnagiri district (excluding hilly areas), Salem, Namakkal district (excluding Tiruchengode taluk) and Perambalur taluk of Perambulur district. The climate prevailing in this region is dry and sub humid. This region has been identified as moderately drought prone area. The elevation varies from 330 to 1070 m above mean sea level. The mean annual rainfall of this region is appreciably lower than in North Eastern zone and is 825 mm received in 47 rainy days. The region is benefited by both south-west and north-east monsoon rains but unlike the NEZ, the former contributed more to the total rainfall. The mean monthly maximum temperature ranges between 30 to 37°C with minimum temperature ranging between 19 and 25.5°C. The annual PET of this region is 1727 mm compared to the annual precipitation of 825 mm.

3. Western zone

This zone comprises of Erode, Coimbatore, Dindugal, Theni districts, Tiruchengode taluk of Namakkal district, Karur taluk of Karur district and some western part of Madurai district. The mean annual rainfall is 718 mm in 45 rainy days. The monthly mean maximum temperature is 35°C in April and 30°C in January and November. The monthly mean minimum temperature is 19°C in January and 24°C in May.

4. Cauvery Delta zone

This zone comprises the Cauvery Delta area in Thanjavur, Thiruvarur, Nagapattinam districts and Musiri, Tiruchirapalli, Lalgudi, Thuraiyur and Kulithalai taluks of Tiruchirapalli district, Aranthangi taluk of Pudukottai district and Chidambaram and Kattumannarkoil taluks of Cuddalore district. The mean annual rainfall of the zone is 1078 mm out of which 40mm is received during winter, 69.2mm during summer, 295.4mm during South West Monsoon and 673.8mm during North East Monsoon.

5. Southern zone

This is the biggest among the seven zones of Tamil Nadu. It is typical zone surrounded by coastal areas on the East and mountains in the West. This zone comprises Sivagangai, Ramanathapuram, Virudunagar, Tuticorin and Tirunelveli districts and Natham and Dindigul taluks of Dindigul district, Melur, Thirumangalam, Madurai South and Madurai North taluks of Madurai district and Pudukkottai district excluding Aranthangi taluk. This zone lies on the Southern part of the State under rain shadow area. Because of this, the area is prone to drought very often. The climate is semi-arid tropics. The elevation varies from mean sea level to 300 m. The mean annual rainfall is 776 mm received in 43 rainy days. The monthly mean maximum temperature in this region ranges from 28.5°C in December to 38.5°C in June. The monthly mean minimum temperature varies from 21.0°C in January to 27.5°C in June.

6. High Rainfall zone

This zone consists of Kaniyakumari district. This district situated in the southern most part of the Peninsular India, with its high rainfall having a climate which is entirely different from the rest of the state. The climate is monsoon tropics and there is seasonal in shores flow of moist air. The elevation ranges from sea level to about 600 m. The mean annual rainfall of the district is 1469 mm received in 64 rainy days. There is not much fluctuation in the mean monthly air temperature. The monthly mean maximum temperature varies from 28.0°C in December to 33.5°C in May. The monthly mean minimum temperature varies from 22°C in December to 26.5°C in May.

7. High altitude and Hilly zone

This zone comprises the hilly regions, namely the Nilgiris, Shevroys, Elagiri-Javvadhu, Kollimalai, Patchaimalai, Anamalais, Palanis and Podhigai malais. The rainfall varies from 850 mm in Kalrayan hills to about 4500 mm in Anamalai hills.

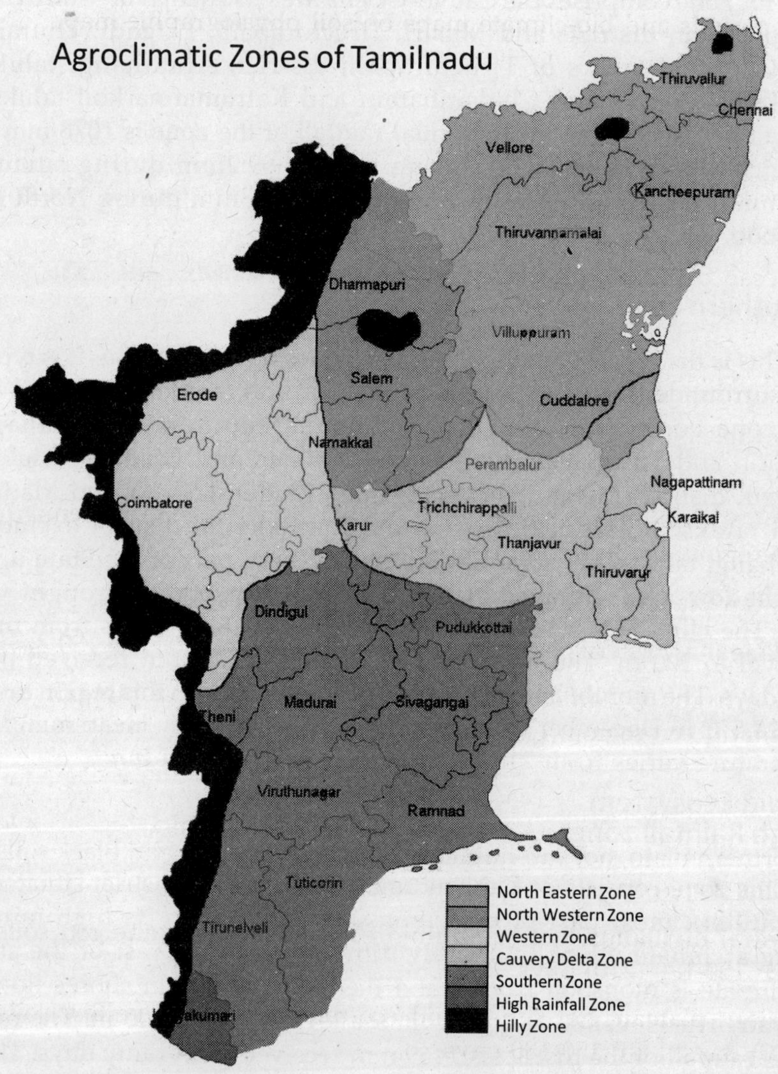

Agro Ecological Zones of India

An ecological region is characterized by distinct ecological responses to macroclimate as expressed in vegetation and reflected in soils, fauna and aquatic systems. Therefore, an agro-ecological region is the land unit on the earth's surface carved out of agro-climatic region when superimposed on different landform and soil conditions that act as modifiers of climate and length of growing period (LGP).

National Bureau of Soil Survey and Land Use Planning (NBSS & LUP) of the ICAR has delineated 20 agro-ecological regions (AERs) in the country using the FAO 1978 concept of superimposition of length of growing periods and bio-climate maps on soil physiographic map.

Arid ecosystem

1. Western Himalayas, cold eco-region, shallow soils, LGP <90 days.
2. Western plain Kachohh and parts of Kathiawar Peninsula, hot arid eco-region, desert and saline soils. LPG <90 days.
3. Deccan plateau, hot arid ecoregion, red and black soils. LGP <90 days.

Semiarid ecosystem

4. Northern plain and central high lands, hot semiarid eco-region, alluvial soils. LGP 90-150 days.
5. Central high lands, Gujarat plains, Kathiawar peninsula, hot semiarid eco-region, medium and deep black soils. LGP 90-150 days.
6. Deccan plateau, hot semiarid ecoregion. LGP 90-150 days.
7. Telangana, Eastern ghats, hot semiarid eco-region. LGP 90- 150 days.
8. Eastern ghats, Tamil Nadu uplands and Karnataka plateau, hot semiarid eco-region. LGP 90-150 days.

Subhumid Ecosystem

9. Northern plain, hot sub-humid (dry) eco-region, red and black soils. LGP 150-180 days.
10. Central highlands, hot sub-humid eco-region, black and red soils. LGP 150-180 (210) days.
11. Eastern plateau, hot sub-humid eco-region, red and yellow soils, (210) days. LGP 150-180 days.

12. Eastern plateau (Chotanagpur) and Eastern ghats hot sub-humid eco-region, red and lateritic soils. LGP 150-180 (210) days.
13. Eastern plain, hot sub-humid (moist) eco-region, alluvial soils. LGP 180-210 days.
14. Western Himalayas, warm sub-humid to humid eco-region with brown forest soils. LGP 180-210+ days.

Humid-Perhumid ecosystem

1. Bengal and Assam plain hot sub-humid (moist) to humid eco-region, alluvial soils. LGP 210+ days.
2. Eastern Himalayas, warm per-humid eco-region, brown and red hill soils. LGP 210 + days.
3. North eastern hills, warm per-humid eco-region, red and lateritic soils. LGP 210+ days.

Coastal Ecosystem

4. Eastern coastal plain, hot sub-humid to semiarid eco-region, coastal alluvium. LGP 90-210 +days.
5. Western Ghats and coastal plain, hot humid-per-humid eco-region, red, lateritic and alluvium derived soils. LGP 210+ days.

Island Ecosystem

6. Andaman Nicobar and Lakshadweep, hot humid to per-humid eco-region, real loamy and sandy soils. LGP 210+ days.

The major advantage of LGP based criteria is that the LGP is the direct indicative of moisture availability of a given landform rather than the total rainfall. For example, both Ratnagiri in western Maharashtra and Nagpur in eastern Maharashtra have LGP 180-210 + days but the total annual rainfall of Ratnagiri is more than 2000 mm where as that of Nagpur is only 1100 mm. Therefore, agro-ecosystems approach allows crop planning based on length of growing period rather than the quantity of rainfall.

Questions

Fill the blanks

1. Agronomy is derived from Greek word __________ meaning 'field' and __________ meaning 'management'.
2. Planning commission of India classified India into __________ agroclimatic zones
3. The agro-ecosystems approach allows crop planning based on __________ rather than the quantity of rainfall.
4. The mean annual rainfall of high rainfall zone is __________ mm.
5. According to NARP classification, Tamil Nadu is classified into __________ agroclimatic zones.

Choose the correct answer

6. The number of agro ecological zones in India is

   a. 7 b. 15

   c. 12 d. 20

7. Agronomist is concerned to meet the needs of growing population with production of

   a. Food and fibre b. Fodder

   c. Sugar crops d. Trees

8. The scientist responsible for every decision made in the farm management is

   a. Agronomist b. Economist

   c. Sociologist d. Pathologist

9. The biggest zone among the seven agroclimatic zones of Tamil Nadu is

   a. Western zone b. Southern zone

   c. Hilly zone d. North western zone

10. In high rainfall zone, the rainfall varies from

    a. 850-4500 mm b. 600-1500 mm

    c. 5000-6500 mm d. 500-800 mm

Answer the following

1. Agronomy
2. Scope of Agronomy
3. Agroclimatic zones of India
4. Agroecological zones
5. Cauvery delta zone.

4

Crops and Cropping System

In general crop is an organism grown or harvested for obtaining yield. Agronomically crop is a plants cultivated for economic purpose.

Classification of Crops

Classification is done to generalize similar crop plants as a class for attaining better understanding of them. Field crops are classified in the following ways.

1. According to range of cultivation
2. According to the place of origin
3. Botanical classification
4. Commercial classification
5. Economic / Agricultural / Agrarian classification
6. Seasonal classification
7. Classification based on ontogeny
8. According to cultural requirement
9. According to important uses

1. Range of Cultivation

Garden crop: Grown on a small scale in gardens. Example – Onion, Brinjal etc.

Plantation crop: Grown on a large scale in estates and perennial in nature. Example – Tea, Coffee, Cocoa, Rubber etc.

Field crop: Grown on a vast scale under field condition. They are mostly seasonal such as rice, wheat, cotton etc. Agronomy deals with field crops only.

2. Place of Origin

a) **Native**: Crops grown within the geographical limits of their origin, for example – rice, barley, blackgram, green gram, mustard, castor, sugarcane and cotton, grown in India, are native to India.

b) **Exotic or Introduced**: Crops introduced from other countries, such as Tobacco, potato, jute, maize, apple, etc.

3. Botanical / Taxonomical Classification

According to systematic botany plants are classified as order, family etc. Similarly, crop plants are grouped into families as,

a)	Poaceae (Graminae)	: Cereals, millets and grasses.
b)	Papilionaceae (Legumes)	: Pulses, legume fodders, vegetables, groundnut, berseem, green manures etc.
c)	Cruciferae	: Mustard, Indian rape seed, radish cabbage, cauliflower
d)	Cucurbitaceae	: All gourds, cucumber, pumpkin etc.
e)	Malvaceae	: Cotton, lady's finger, rosselle
f)	Solanaceae	: Potato, tomato, tobacco, chillies, brinjal
g)	Tiliaceae	: Jute
h)	Asteraceae (Compositae)	: Sunflower, safflower, ginger
i)	Chenopodiaceae	: Spinach, beet, sugarbeet.
j)	Pedeliaceae	: Sesame
k)	Euphorbiaceae	: Castor, tapioca
l)	Convolvulaceae	: Sweet potato
m)	Umbelliferae	: Coriander, cumin, carrot, anise
n)	Liliaceae	: Onion, garlic
o)	Zingiberaceae	: Ginger, turmeric

4. Commercial classification

Based on the plant products which come into the commercial field are grouped as

a) **Food crops**: Rice, wheat, greengram, soybean, groundnut, etc.

b) **Food crops / Forage crops** – All fodders, oats, sorghum, maize, napier grass, stylo, Lucerne etc.

c) **Industrial / Commercial crops**: Cotton, sugarcane, sugarbeet, tobacco, jute, etc.

d) **Food adjuncts**: Turmeric, garlic, cumin, etc.

5. Economic / Agrarian / Agricultural Classification

This classification is based on use of crop plants and their products. This is an important classification as for as agronomy is concerned (*Agronomic classification)* (For Botanical names of crops see **annexure - III**)

a) **Cereals**: They are cultivated grasses grown for their edible starchy grains (one seeded fruit - caryopsis) Larger grains used as staple food are cereals - rice, wheat, maize, barley, oats etc. The word cereal was derived from the word ceres which denotes a goddess who was believed as the giver of grains by Romans.

b) **Millets:** Small grained cereals which form the staple food in drier regions of the developing countries are called millets. Example - sorghum, bajra, ragi and minor millets.

c) **Oilseeds**: Crops that yield seeds that are rich in fatty acids, are used to extract vegetable oils. Example - mustard, rape seed, sesame, sunflower, castor, linseed, groundnut, soybeans etc.

d) **Pulses**: Seeds of leguminous plants used as food. They produce dal rich in protein. Examples - greengram, blackgram, redgram, bengal gram, soybean, cowpea, peas, lentil, lathyrus.

e) **Feed / Forage**: It refers to vegetative matter, fresh or preserved, utilized as feed for animals. It includes hay, silage, soilage, pasturage and fodder. Example - bajra napier grass, guinea grass, fodder sorghum, fodder maize, lucerne, desmanthus, etc.

f) **Fibre crops**: Plants grown for their fibre yield. There are different kinds of fibre. They are i) seed fibre - cotton, ii) stem/bast fibre- jute, mesta, iii) leaf fibre - agave, pineapple.

g) **Sugar and starch crops**: Crops grown for production of sugar and starch. Example - sugarcane, sugarbeet, potato, sweet potato, tapioca and asparagus.

h) **Spices and Condiments**: Crop plants or their products used to season, flavour, taste, and add colour to the fresh or preserved food. Example ginger, garlic, fenugreek, cumin, turmeric, chillies, onion, coriander, anise and asafoetida.

i) **Drug crops / Medicinal plants**: Crops used for preparation of medicines. Example - tobacco, mint etc.

j) **Narcotics, fumitories and masticatories**: Plants / products used for stimulating, numbing, drowsing or relishing effects. Example - tobacco, ganja, opium poppy.

k) **Beverages**: Products of crops used for preparation of mild, agreeable and stimulating drinking. Example - tea, coffee, cocoa.

6. Seasonal Classification

Crops are grouped under the seasons in which their major field duration falls.

a) ***Kharif* or South West Monsoon season crops:** Crops grown during June - July to September - October which requires a warm wet weather during their major period of growth and shorter day length for flowering. Example - rice, maize, castor, groundnut,

b) ***Rabi* Crops:** Crops grown during October - November to January - February which require cold dry weather for their major growth period and longer day length for flowering. Example - wheat, mustard, barley, oats, potato, bengal gram, berseem, cabbage and cauliflower.

c) ***Zaid* or summer crops**: Crops grown during February - March to May - June which requires warm dry weather for growth and longer day - length for flowering. Example - blackgram, greengram, sesame, cowpea etc.

This classification is not a universal one. It only indicates the period when a particular crop is raised. Example - khairf rice, kharif maize, rabi, maize, summer pulse etc.

7. According to Ontogeny

It is a classification based on the life cycle of a plant.

a) **Annual Crops**: Crop plants that complete life cycle within a season for year. They produce seed and die within the season. Example- wheat, rice, maize, mustard.

b) **Biennial Crops:** Plants that have life span of two consecutive seasons or years. First year / Season these plants have purely vegetative growth usually confined to rosette of leaves. The tap root is often fleshy and serves as a food storage organ. During the second year/

season they produce flower stocks from the crown and after producing seeds the plants die Example-sugarbeet, beet root, cabbage, radish, carrot, etc.

c) **Perennial Crops**: They live for three or more years. They may be seed bearing or non-seed bearing. Example – sugarcane, napier grass. In general perennial crops occupy land for more than 30 months.

8. According to cultural requirement of crops

Certain groups of plants are alike in cultural requirements due to their similar agro-botanical or morpho – agronomical characters.

a. According to suitability of toposequence

i) **Crops grown on high land (dry land crops)**: Crops that cannot tolerant water stagnation come under this group example redgram, groundnut, maize, sorghum, cotton, sesame, napier etc.

ii) **Crops grown on medium land (garden land crops):** Crops that require sufficient soil moisture but cannot tolerate water stagnation. Example – Potato, sugarcane, upland rice, ragi, wheat, blackgram, bengal gram.

iii) **Crops grown on low land (Wetland crops)**: Crops that require abundant supply of water and can withstand prolonged water logged conditions. Example – rice, daincha, paragrass and jute.

b. According to the suitability of the textural groups of soils.

i) **Crops suitable to sandy to sandyloam (light) soils** : Sorghum, bajra, green gram, sunflower, potato, onion, carrot etc.

ii) **Crops suitable to silty loam (medium) soils** : Jute, sugarcane, maize, cotton, mustard, tobacco, bengal gram, redgram, cowpea, etc.

iii) **Crop suitable to clay to clay loam (heavy) soils** : Rice, wheat, barley, linseed, lentil, paragrass, guinea grass, marvel grass etc.

c. According to tolerance to problem soils,

i) **Tolerant to acidic soils:** Wet rice, potato, mustard.

ii) **Tolerant to saline soils:** Chillies, cucurbits, wheat, sorghum, bajra, cluster beans, barley etc.

iii) **Tolerant to alkali / sodic soils:** Barley, cotton, bengalgram, berseem, sunflower, maize, etc.

iv) **Tolerant to waterlogged soils:** Wet rice, daincha, paragrass, napier grass, guinea grass.

v) **Crops tolerant to soil erosion:** Marvel grass, groundnut, blackgram, rice bean, moth bean, horsegram.

d. According to tillage requirement

i) **Arable crops** : require preparatory tillage. Example – Potato, tobacco, rice, maize.

ii) **Non-arable crops** : may not require preparatory cultivation / tillage. Example – Blackgram, green gram, Paragrass, cowpea.

e. According to the depth of root system

i) **Shallow rooted crops :** Rice, potato, onion

ii) **Moderately deep rooted :** Wheat, groundnut, castor, tobacco.

iii) **Deep rooted :** Maize, cotton, sorghum

iv) **Very deep rooted :** Sugarcane, safflower, lucerne, redgram.

f. According to the tolerance to hazardous weather condition

i) **Frost tolerant :** Sugarbeet, beet root,

ii) **Cold tolerant :** Potato, cabbage, mustard

iii) **Drought tolerant :** Bajra, jowar, barley, safflower, castor

g. According to water supply

i) **Irrigated crops :** Rice, wheat, pulses, groundnut, berseem.

ii) **Rainfed / uplant crops :** Jute, maize, ragi, upland rice, redgram

iii) **Rainfed but partially irrigated :** Bengal gram, wheat, jowar, bajra, mustard

iv) **Residual / conserved soil moisture :** Rapeseed, bajra, barley, safflower, linseed.

v) **Rainfed with supplemental irrigation :** *Kharif* rice, sugarcane, blackgram.

vi) **Rainfed with flooded water :** Deep water rice, jute, sugarcane, daincha.

h. According to method of sowing / planting

i) **Direct seeded crop :** Upland rice, wheat, jowar, bajra, groundnut etc.

ii) **Planted crops :** Sugarcane, potato, sweet potato, napier, guinea grass.

iii) **Transplanted crops** (after raising seedlings in the nursery) : Rice, ragi, bajra, tobacco, bellary onion, brinjal.

i. According to inter-tillage requirements specially earthing up

i) **Intertilled crops : Potato,** sweet potato, groundnut, maize, sugarcane, turmeric, etc.

ii) **Non-intertilled crops :** Fodder sorghum, deenanath grass, paragrass etc.

j. According to length of field duration of crops

i)	Very short duration crops	: upto 75 days	: Pulses
ii)	Short duration crops	: 75–100 days	: Sunflower, cauliflower, uplandrice
iii)	Medium duration crops	: 100-125 days	: Wheat, jowar, bajra,groundnut, sesame, jute.
iv)	Long duration crops	: 125-150 days	: Mustard, tobacco, cotton
v)	Very long duration crops	: above 150 days	: Sugarcane, redgram, castor.

k. According to the method of harvesting

i)	Reaping	: Rice, wheat
ii)	Uprooting by pulling	: Bengal gram, black gram, lentil, rapeseed
iii)	Uprooting by digging	: Potato,sweetpotato, groundnut, carrot etc.
iv)	Picking	: Cotton, vegetables, brinjal, bhendi, chillies
v)	Priming	: Tobacco
vi)	Cutting	: Berseem, napier, amaranthus
vii)	Grazing	: Paragrass, kolukkattai grass, stylo.

l. According to post harvest requirement

i)	Curing	: Tobacco, mustard
ii)	Stripping	: Jute, sunnhemp
iii)	Shelling	: Groundnut
iv)	Ginning	: Cotton
v)	Seasoning	: Turmeric, chillies
vi)	Grading and sorting	: Potato, rice, wheat, fibre crops etc.

m. Based on crops growing soil condition

i)	Psammophytes (Sandy soil)	: Castor
ii)	Lithophytes (Rock surface)	: Ferns
iii)	Chasmophytes (Rock crack)	: Potato
iv)	Acedophytes (Acid soil)	: Potato
v)	Basophytes (Alkali soil)	: Rice
vi)	Calciphytes (Basic soil)	: Asparagus
vii)	Halophytes (Saline soil)	: Sugar beet, Alfalfa

n. Based on climatic condition

i)	Tropical crop	: Coconut, sugarcane
ii)	Sub-tropical crop	: Rice, cotton
iii)	Temperate crop	: Wheat, barley
iv)	Polar crop	: All pines, pasture grasses

9. According to important uses

Though plants are useful in many ways only certain uses are given below :

a) **Catch crops / contingent crops:** are those crops cultivated to catch the forth coming season. It replaces the main crop that has failed due to biotic or climatic or management hazards. Generally, they are of very short duration, quick growing, fast bulking, harvestable or usable at any time of their field duration and adaptable to the season, soil and management practices. They provide feed, check weed growth, conserve soil, utilized added fertilizer and moisture Example – green gram, black gram, cowpea, onion, coriander, bajra.

b) **Restorative crops:** are those crops which provide a good yield along with enrichment or restoration of soil fertility or amelioration of the soils. They fix atmospheric nitrogen in root nodules, shed their leaves during ripening and thus restore soil conditions. Example - legumes.

c) **Exhaustive crops:** are those crop plants which on growing leave the field exhausted because of a more aggressive nature. Example - gingelly, brinjal, linseed, sunflower etc.

d) **Pair crop / residual crops:** are those crop plants which are sown a few days or weeks before the harvest of the standing mature crops to utilize the residual moisture, without preparatory tillage. The standing crop and the later sown (pair) crop become simultaneous (forming a pair) for a short period. For example, Black gram in thaladi paddy; lathyrus, lentil etc., in paddy, Paira crops in succession may constitute relay cropping.

e) **Smother crops:** are those crop plants which are able to smother or suppress the weed growth by providing suffocation (curtailing movement of air) and obscuration (of the incidental radiation) through their dense foliage developed due to quick growing ability with heavy tillering or branching, planophyllic or procumbent or trailing habits. Example - barley, mustard, cowpea, etc.

f) **Cover crops:** are those crop plants which are able to protect the soil surface from erosion (wind, water or both) through their ground covering foliage and or root mats. Example - groundnut, blackgram, marvel grass, sweet potato.

g) **Nurse crops:** A companion crop which nourishes the main crop by way of nitrogen fixation and or adding the organic matter into the soil. Example-cowpea intercropped with cereals, *glyricidia, tephrosia* in tea.

h) **Guard / barrier crops:** are those crop plants which help to protect another crop from trespassing or restrict the speed of wind and thus prevent crop damage. Main crop in the centre is surrounded by hardy or thorny crop. Example -mesta around sugarcane; sorghum around cotton; safflower around gram.

i) **Trap crops:** are those crop plants grown to trap soil borne harmful parasitic weeds. For example orobanche and striga are trapped by solanaceous and sorghum crops respectively. Nematodes are trapped by solanaceous crops (On uprooting crop plants, nematodes are removed from the soil). Castor in cotton, groundnut act as trap crop for army worm pest.

j) **Augmenting crops:** are those sub crops sown to supplement the yield of the main crop. Example – Mustard or cabbage with berseem to augument the forage yield of berseem.

k) **Alley crops:** are those arable crops which are grown in 'alleys' formed by trees or shrubs, established mainly to hasten soil fertility restoration, enhance soil productivity and reduce soil erosion. They are generally of non-trailing with shade tolerance capacity. For example growing pulses in between the rows of casuarina.

Crop Adaptation and Distribution

Adaptation may be defined as any feature of an organism which has survival value under the existing condition of its habitat. Such features or feature may allow the plants to make fuller use of nutrients, water, temperature or light available or may give protection against adverse factors such as temperature extremes, harmful insects and diseases. Adaptation may be morphological or physiological.

Morphological adaptation such as growth habit, strength of stalk, radial symmetry, or rhizomes.

Physiological adaptation which result in resistance to parasites, greater ability to compete for nutrients or ability to withstand desiccation. However both morphological and physiological adaptation represents the expression of physiological processes.

Principles of Plant distribution: Environmental factors are highly influential in determining the natural distribution of plants.

There are eight principles of plant distribution

1. Evolution
2. Climatic factors like light, temperature, moisture, wind etc.,
3. Edaphic factors like soil, parent material, physiography
4. Dispersal of flora
5. Plant migrations
6. Climatic variations or change
7. Relative distributions of land and sea (occurrence in geological time) and it exerts a high degree of control over distribution of flora
8. Biotic factors like obligate insect pollination, seed dissemination by animals and grazing by livestock directly influence plant distribution.

Theories Governing Crop Adaptation and Distribution

1. **Theory of tolerance**: Each plant or living organisms is able to thrive well in certain climatic conditions below which and above which the plant can't grow. i.e. it requires optimum climatic conditions. Temperature is one of the most common limiting factors in plant distribution. Many tropical crops such as rubber, cocoa, banana will not with stand freezing temperature (0^0C) In these rubber probably has the narrowest tolerance range and banana the widest range for temperature tolerance.

2. **Theory of avoidance**: It may be accomplished through rapid completion of the life cycle, as in ephimerals, dormancy in seeds to avoid effects of the hottest and driest periods, dormancy in vegetative parts or roots of all the perennials, water accumulation in succulents and extremely deep root systems to avoid moisture deficiency.

3. **Theory of Factors replaceability:** One factor that can be replaced by another or substituted by another. For example,

   a) Elevation can be substituted for latitude because of its temperature effects. The climatic conditions at the latitudes of 35 - 45^0 N resembles to that of tropical regions at elevation of 4000-6000 ft.

   b) The angle direction of slope may be substituted for latitude. This is also a temperature adjustment, depending on the angle of exposure to solar radiation, wind etc.

   c) Parent materials may compensate for climate.

   d) Rainfall may be replaced by fog and to some extent by dew

   e) Soil texture may be substituted for moisture.

Major Crops of India

1. Rice

In India rice is the most important food crop and it is the staple food in tropical and subtropical regions of Asia and Africa.

Origin: Indo-Burma (Indo-Myanmar)

Adaptation: Grown in the world between 39^0S (Australia) 50^0N latitude (China). In India it is grown between 8^0 N to 34^0N latitude.

Altitude: From below sea level (Kuttanad region of Kerala) to 3000 m (Jammu and Kashmir) above MSL.

Rainfall and Temperature: Rice is classified as a hydrophyte. A heavy rainfall (R.F) of 125 cm is required during its growing period. There should be a monthly R.F. 0f 200 mm to grow lowland rice and 100 mm to grow upland rice respectively. Deep water rice requires one meter height of standing column of water. Rice requires high humidity and high temperature (18-32^{0}C). The critical mean temperature for flowering and fertilization is 16-20^{0}C.

Soil: Though rice can be grown in variety of soils, ideal soil is heavy alluvial soils of river valley and delta. The best is soils with slightly acidic nature 5.5 to 6.5 pH, but rice is commonly grown in soils of 4.5 to 8.5 pH. Rice is also grown in acidic peaty soils of kerala with pH of 3.0. and highly alkaline soils of Punjab and Hariyana with pH 10.0.

Distribution: Rice is widely distributed in India, Pakistan, Bangaladesh, Malaysia, Taiwan, China, Japan, Australia, USA, Spain, Korea.

In India rice is grown in the states of Tamil Nadu, Kerala, Bihar, Uttar Pradesh, Madhya Pradesh, West Bengal, Orissa, Andhra Pradesh etc.

2. Wheat

Wheat is the most important and widely cultivated crop in the world. In occupies a prime position in terms of production. India ranks second in production next to china. In India, wheat is the second most important food crop next to rice.

Origin: Central Asia.

Cultivated species

1. Common / bread wheat - 95% production.
2. Durum / macaroni / samba wheat - 3-4 % production.
3. Emmer wheat - 1% production
4. Indian dwarf wheat - less than 1% production

Distribution: Widely distributed in USSR, China, USA, Swizerland, France, Germany, India etc. In India wheat is grown in the states of Uttar Pradesh, Madhya Pradesh, Punjab, Rajasthan, Bihar, Haryana, Maharashtra and Gujarat.

Adaptation: It can be cultivated from sea level to as high as 3,300 m above MSL.

Climate: Cool winter and dry hot summer is required. Wheat requires a rainfall of 40-90 cm. But high temperature and high humidity are harmful.

Soils: Though grown in wide range of soil, well drained loams and clay loams are better suited. It is grown in soils with pH above 5.8 and the most suitable pH is 6.5 - 7.5

Seasons

Winter wheat - Long duration wheat varieties are grown in this season which requires low temperature during early growth for flowering and fruiting. It is grown from October, November to May, July

Spring wheat: These varieties do not require low temperature for flowering and fruiting. Normally grown from March, May to August, September.

In India, spring wheat is grown in winter (October, November to March, April). There are two seasons for wheat in hills of Tamilnadu (i) October - April and (ii) May - September.

Wheat zones of India

1. North Western plains,
2. North eastern plains,
3. Central zone,
4. Peninsular zone.

3. Maize

It is grown in USA, Brazil, China, Mexico, India and Canada. In India it is grown in states of Uttar Pradesh, Madhya Pradesh, Bihar, Rajasthan, Punjab, Karnataka, Himachal Pradesh. It requires a mean temperature of 24⁰C and night temperature more than 15⁰C. Summer temperature below 19⁰C is not suitable. It requires a well distributed rainfall of 50-75 cm. It can be grown from sea level to 3000 m above MSL.

4. Sorghum

It is grown in USA, China, Nigeria, Sudan and Argentina. In India, it is grown in states of Maharashtra, Andhra Pradesh, Karnataka, Gujarat, Madhya Pradesh, Tamilnadu, Rajasthan and Uttar Pradesh. The temperature requirement is minimum 8 to 10^{0}C, optimum 26-29^{0}C and maximum 35-40°C. Sorghum can tolerate high temperature throughout its life cycle better than any other cereal crops. Sorghum can tolerate drought condition. Because a) it remains dormant during moisture stress and resumes growth when favourable condition reappear and b) it possesses i) high resistance to desiccation, ii) low transpiration rate and iii) largest number of fibrous roots.

5. Bengal gram

It is grown in India, Pakistan, Ethiopia, Burma, Turkey. In India, it is grown in states of Madhya Pradesh, Uttar Pradesh, Rajasthan, Punjab, Haryana and Maharashtra. It is a winter season crop but severe cold and frost are injurious to it. It requires a moderate rainfall of 60-90 cm.

6. Pigeon pea (Arhar)

It is the second most important pulse crop of Inida and foremost in Southern India. It is crop with great resilience and withstands water stress and association of short duration crops without any considerable adverse effect on yield. It is always grown with sorghum / bajra in Tamilnadu.

7. Groundnut

It is a tropical crop grown in India, China, U.S.A. and Brazil. In India, it is grown in the states of Gujarat, Andhra Pradesh, Tamil Nadu and Punjab. It is grown between 45^{0}N and 30^{0}S latitude with R.F:370-600 mm, minimum temperature: 14-16^{0}C and optimum temperature : 21-26.5^{0}C.

8. Sunflower

It requires a rainfall of 380 mm in summer and 550 mm in sandy loam soils. Sunflower has a helio tropic response. It is grown in U.S.A, U.S.S.R, Argentina, Romania, Spain, Yugoslovakia, Turkey and USA. Being thermo and photo insensitive it can be grown throughout the year.

9. Mustard

It is grown in India, China, Pakistan and Bangaladesh. In India, it is grown in Uttar Pradesh, Madhya Pradesh, Rajastan, Punjab, Haryana, Bihar, Orissa, West Bengal and Gujarat. It requires cool climate and rainfall of 35-45 cm.

10. Cotton

It is grown in India, U.S.A, USSR, China, Brazil, Egypt, Pakistan, Mexico, Turkey and Sudan. In India, it is grown in Maharashtra, Gujarat, Karnataka, Andhra Pradesh, Tamil Nadu, Punjab, Rajasthan, Haryana and Uttar Pradesh. Cotton is a heat loving plant requires a minimum rainfall of 175-200 mm (well distributed). It requires a minimum of 180-200 frost free days.

11. Jute

It is grown in Inida, Bangaladesh, China, Thailand, Brazil, Peru, Burma, Nepal and Vietnam. In India, it is grown in West Bengal, Bihar, Orissa, Uttar Pradesh, Mehalaya and Tripura. It requires an optimum temperature of 25-38^{0}C, rainfall of 150 cm/annum and relative humidity of 55-90%.

12. Tobacco

It is grown in India, China and USA. In India, it is grown in Andhra Pradesh, Gujarat, Karnataka, Tamil Nadu, Orissa, West Bengal, Bihar, Maharashtra and Uttar Pradesh. It is a day neutral plant. It requires a rainfall of 500 mm and 90-120 frost free days. The temperature requirements are, minimum temperature 13-14^{0}C, Optimum temperature 27-32^{0}C and maximum temperature 35^{0}C.

13. Sugarcane

Grown in India, Cuba, Brazil, Mexico, Pakistan, China, Philippines, Thailand and USA. In India, it is grown in the states of Andhra Pradesh, Gujarat, Orissa, West Bengal, Bihar, Maharashtra, Karnataka and Tamilnadu. (It is grown in 30^{0}N - 30^{0}S latitude) Frost causes injury to sugarcane buds. It requires an annual rainfall of 1250-2500 mm. It is a short day plant, flowering can be photoperiodically controlled. It requires an optimum temperature of 26-32^{0}C for growth.

14. Potato

Being a crop of temperate regions requires a cool temperature. Ideal temperature for vegetative growth is 24°C and that for tuberisation is 18-20°C. It is susceptible to frost and requires bright sunny weather. High humidity coupled with cloudy days is injurious to potato because the crop is attacked by fungal diseases (late blight).

15. Sugarbeet

It is grown in USSR., USA., France, Germany, Italy, Turkey, Poland, Czechoslovakia. In India, it is grown in Jammu and Kashmir, Punjab, Rajasthan, Uttar Pradesh and Maharashtra, It requires a minimum temperature of °C, optimum temperature of 20 - 22°C and maximum temperature of 30°C and the crop is highly tolerant to frost and cold.

16. Banana

It is grown in India, Taiwan, Equador, Coasta rica, Panama, Mexico, Ivory çoast, Columba and Guatimala. In India, it is grown in Tamilnadu and Kerala. It is grown with rain fall of 1800-2500 mm. It requires a minimum temperature of 8-9°C and optimum temperature of 24-29°C.

Factors Governing Choice of Crop and Varieties

(i) Climate

a) Seasons

Kharif Season Crops: Rice, Maize, Sorghum, Bajra, Ragi, Minor Millets.

Rabi Season Crops: Wheat, Barley, Oats, Chickpea, Sorghum, Potato, Rape Seed, Mustard and Safflower.

Summer Season Crops: Gingelly, Blackgram, Greengram

b) Rainfall

(i) > 30cm/month for atleast 3 months	Rice
(ii) 20-30cm/month for not less than 3 months	Maize/Blackgram
(iii) 10-20cm/month for at least 3 months	Bajra, Small millets
(iv) Rainfall 5-10cm/month	Grasses
(v) < 5cm	Not suited for crop production

(c) Length of crop season

Effective crop growing period	Cropping system
< 20 weeks	Sole crop
20-30 weeks	Sole crop + Inter crop
> 30 weeks	Two crops in sequence

(ii) Natural Resources

a) **Soil:** Cropping pattern is governed by rainfall and soil characteristics.

Rainfall (mm)	Moisture holding capacity of soil (mm)	Cropping pattern
<350	-	Not suited for crop production
350-650	100	Single crop
650-750	100	Intercropping
750-900	150	Sequential cropping (Relay cropping)
>900	200	Double cropping assured

b) **Water**: According to water release in the canal it is decided whether single (late release of water) or double crop (early release of water) can be grown in a year. Depending upon the availability of water from river, canal or well source the crops are selected based on their water requirement.

For example the following is the water requirement of some crops; Banana 2000-2200 mm; Rice 1100-1150 mm; Sorghum 400-450 mm; Cotton 550-650 mm.

(iii) Socio-Economic aspects of farmer

Big farmers and rich farmers can purchase inputs like fertilizers, pesticides and apply in time, but poor small and marginal farmers cannot do so. Education status, knowledge about principles and practices of crop production, technological know-how and skill also plays a major role in crop selection and management.

(iv) Marketing facilities

Mostly marketing facilities are available near the town, cities and agro based processing industries. Farmers prefer a crop, the produce of which will fetch high price in the market. In earlier days sunflower and soybean were grown less due to lack of processing industry. Now, due to industrial development they are grown in large scale.

(v) Economics

The ultimate objective of commercial farming is to produce more produce per rupee invested. Based on this criteria farmers select the high yielding varieties of maize, sorghum, sunflower and hybrid maize, hybrid cotton which produce more yield than local varieties. Generally farmers are interested to grow the hybrids to get maximum monetary benefits i.e., more income per rupee invested.

Cropping Pattern and Cropping System

Traditionally, increased food production has come from putting more land under cultivation. However, in large areas of the world, especially in Asia, all the land that can be economically cultivated is already in use. In future, most of the extra food needs must come from higher production from land already being farmed. A major share of this increase is likely to come from increasing the number of crops produced per year on a given land using improved crop cultivars. Such multiple cropping offers potential not only to increase food production but also land degradation.

In India, the concept of cropping systems is as old as agriculture. Farmers preferred mixed cropping, especially under dry land conditions, to minimise the risk of total crop failure. Even in Vedas, there is a mention of first and second crops, indicating the existence of sequential cropping.

A **system** is defined as a set of components that are interrelated and interact among themselves.

A **cropping system** refers to a set of crop systems, making up the cropping activities of a farm system.

Cropping system comprises all components required for the production of a particular crop and the interrelationships between them and environment (TAC, CGIAR, 1978). In other words, a cropping system usually refers to a combination of crops in time and space. Combination in time occurs when crops occupy different growing period and combinations in space occur when crops are inter planted. When annual crops are considered, a cropping system usually means the combination of crops within a given year.

Cropping pattern

The yearly sequence and spatial arrangement of crops or of crops and fallow on a given area.

Cropping system

The cropping patterns used on a farm and their interaction with farm resources, other farm enterprises, and available technology which determine their make up.

Intensive Cropping

Principles

- The turn around period between one crop and another is minimised through modified land preparation.
- It is possible when the resources are available in plenty. Ex. Garden land cultivation.
- Cropping intensity is higher in intensive cropping system.
- Crop intensification technique includes intercropping, relay cropping, sequential cropping, ratoon cropping, etc.
- All such systems come under the general term multiple cropping.

Need for intensive cropping

- Cropping systems has to be evolved based on climate, soil and water availability for efficient use of available natural resources.
- The increase in population has put pressure on land to increase productivity per unit area, unit time and for unit resource used.
- This cropping system should provide enough food for the family, fodder for cattle and generate sufficient cash income for domestic and cultivation expenses.

Intensive cropping: Growing number of crops on the same piece of land during the given period of time.

Cropping intensity: Number of crops cultivated in a piece of land per annum is cropping intensity. In Punjab and Tamil Nadu, the cropping intensity is more than 100% (i.e. around 140-150%). In Rajasthan, the cropping intensity is less.

Multiple cropping: The intensification of cropping in time and space dimensions. Growing two or more crops on the same field in a year.

Forms of multiple cropping

Intercropping: Growing two or more crops simultaneously on the same field. Crop intensification is in both time and space dimensions. There is intercrop competition during all or part of crop growth.

a) *Mixed intercropping:* Growing two or more crops simultaneously with no distinct row arrangement. Also referred to as mixed cropping. Ex: Sorghum, pearl millet and cowpea are mixed and broadcasted in rainfed conditions.

b) *Row intercropping:* Growing two or more crops simultaneously where one or more crops are planted in rows. Often simply referred to as intercropping. Maize + greengram (1:1), Maize + blackgram (1:1), Groundnut + Rredgram (6:1)

c) *Strip intercropping:* Growing two or more crops simultaneously in strips wide enough to permit independent cultivation but narrow enough for the crops to interact agronomically. Ex. Groundnut + redgram (6:4) strip.

d) *Relay intercropping:* Growing two or more crops simultaneously during the part of the life cycle of each. A second crop is planted after the first crop has reached its reproductive stage of growth, but, before it is ready for harvest. Often simply referred to as relay cropping. Rice-rice fallow pulse.

Advantages of intercropping

- Better use of growth resources including light, nutrients and water
- Suppression of weeds
- Yield stability; even if one crop fails due to unforeseen situations, another crop will yield and gives income
- Successful intercropping gives higher equivalent yields (yield of base crop + yield of intercrop), higher cropping intensity
- Reduced pest and disease incidences
- Improvement of soil health and agro-eco system

Sequential cropping: Growing two or more crops in sequence on the same field in a farming year. The succeeding crop is planted after the preceeding crop has been harvested. Crop intensification is only in time dimension. There is no intercrop competition.

a) *Double, triple and quadruple cropping:* Growing two, three and four crops, respectively, on the same land in a year in sequence.

Ex. Double cropping: Rice: cotton; Triple cropping: Rice: rice: pulses; Quadruple cropping: Tomato: ridge gourd: *Amaranthus* greens: baby corn

b) *Ratoon cropping:* The cultivation of crop re-growth after harvest, although not necessarily for grain. Ex. Sugarcane: ratoon; Sorghum: ratoon (for fodder).

The various terms defined above bring out essentially two underlying principles, that of growing crops simultaneously in mixture, i.e., intercropping; and of growing individual crops in sequence, i.e., sequential cropping. The cropping system for a region or farm may comprise either or both of these two principles.

Difference between Intercropping and Mixed Cropping

Intercropping	Mixed Cropping
1. The main objective of intercropping is to utilize the space between rows of main crop and to produce more grain per unit area	The main objectives of mixed cropping is insurance against crop failure. Purpose is to get atleast one crop under any climatic, disease or insect hazards
2. There is no competition between main and subsidiary crops	There is competition between crops. Here all crops area given equal care and there is no main or subsidiary crop
3. In intercropping, the main crop is of long duration and subsidiary crop short duration / early maturing	Generally crops are of the same duration (and may be of different duration also)
4. Main and subsidiary crops are sown in rows	Mixed cropping, generally crop seed are mixed and broadcasted
5. The sowing time of both the crops may be the same or the main crop is sown earlier than the subsidiary crop	The sowing time for all crops is same

Crop Rotation – Principles and Advantages

If the same crop is repeatedly grown on the same land it is referred as *monoculture* or *monocropping* (Example - Rice-Rice-Rice) whereas *crop rotation* is the repetitive cultivation of an orderly succession of different crops and crops and fallow on the same land. One cycle may take several years to complete (Example - Rice-Rice-Pulse) Sugarcane - Ratoon sugarcane - Rice (2 or 3 years) Banana - Ratoon Banana - Rice (3 years).

Crop rotation: May also be defined as a process of growing different crops in succession on a piece of land in a specific period of time with an object to get maximum profit from minimum investment without impairing the soil fertility.

Advantages of crop rotation

1. Crop rotation helps in maintaining of soil fertility, organic matter content and recycling of plant nutrients. All crops do not require the plant nutrients in the same proportion. If different crops are grown in rotation, the fertility of land is utilized more evenly and effectively.
2. Restorative crops like heavy foliage crops and green manure crops included in rotation increases the nitrogen and organic matter content of the soil.
3. Helps in control of specific weeks like bermuda grass, *Cyprus* (sedges); *Trianthema portulacastrum.*
4. Avoids accumulation of toxins and maintains physical properties of soil.
5. Controls certain soil borne pests and disease.
6. Reduces and spreads the pressure of work due to different farm operations in a stipulated period of time.

Principles of crop rotation

1. The crops with tap roots (deep rooted) should be followed by those which have fibrous (shallow) root system. This helps in proper and uniform use of nutrients from the soil.
2. The leguminous crops should be grown before non-leguminous crops because legumes fix atmospheric N into soil and add more organic matter to the soil.
3. More exhaustive crops should be followed by less exhaustive crops because crops like potato, sugarcane, maize etc., need more inputs such as better tillage, more fertilizers, greater number of irrigations etc.

4. Selection of the crop should be demand based.
5. The crop of the same family should not be grown in succession because they act as alternate hosts for insect pests and diseases.
6. An ideal crop rotation is one which provides maximum employment to the farm family and labour and permits efficient use of machines and equipments and ensures timely agricultural operations simultaneously maintaining soil productivity.
7. The selection of the crops should be problem based i.e.
   i) On sloppy lands which are prone to erosion an alternate cropping of erosion promoting and erosion resisting crops like legumes should be adopted.
   ii) In low-lying and flood prone area, the crops which can tolerate water stagnation should be selected.
   iii) Under dry farming the crops which can tolerate the drought should be selected.
8. The selection of crops should suit farmer's financial conditions.
9. The crop selected should also suit the soil and climatic conditions.

Factors Determining the Cropping System of a Locality or Region

Prevailing or existing crops and varieties in a cropping system is the cumulative results of past and present decisions of individuals, communities and Government and their agencies.These decisions were based on experimentation, tradition, expectation, profit (economics) personal preferences and resources, political and social pressures and so on. The choice of crops and varieties in a particular locally or region is decided by.

i) Climate prevailing in a locality,
ii) Natural resources like soil, water, light and air,
iii) Socio-economic aspects,
iv) Marketing facilities and
v) Economics of the cropping programme.

In general, cropping system is developed taking into account

i) Availability of resources (input) and managerial skill of farmers suggested for each crop or variety and in crop combination,

ii) Ecologically practicable pest and disease control methods to the existing cropping system and the proposed cropping system,

iii) Interaction of the existing and chosen crop or introduced crops individually or in combination,

iv) Economics of a cropping system prevailed in a region,

v) Influence of infrastructure facilities including marketing on the ecologically feasible crops of a regional and

vi) Operational factors deciding the existence of economically preferable crops or cropping system in a region.

To describe the cropping pattern of a region the crop occupying the highest percentage of the sown area in a particular season or year of the region is taken as the base crop.

i) A substitute for base crop in the same season.

ii) Crop which fit in with the rotation in the subsequent seasons

iii) Supplementary crops-which are grown in addition to base crop as intercrops.

Factors Determining the Cropping Pattern

1. Climate

   a) Atmospheric temperature

   b) Occurrence of rainfall:

      i) Quantity of rainfall and

      ii) Period of rainfall.

2. Topography

   Altitude and slope will decide the crops and cropping pattern of a locality.

3. Soils

   a) Black soil - cotton

   b) Red soil - sorghum, redgram, groundnut

   c) Laterite and Lateritic soils - rice/tea

   d) Alluvial soils,

   e) Sierozemic soils,

   f) Chesnut or brown soils.

Cropping Patterns Followed in India

Cropping patterns in India: India may be broadly divided into five agricultural regions.

1. **Rice region** – North East, South West and West Coast regions of India
2. **Wheat region**: North, West and Central India.
3. **Millet region (sorghum)**: Maharashtra, Rajasthan, Gujarat, Madhya Pradesh and Deccan Plateau, Southern peninsula.
4. **Temperate Himalayan region**: Kashmir, Himachal Pradesh, Uttar Pradesh.
5. **Plantation crops region**: Assam and hills of South India.

A) Kharif season cropping patterns

1. Rice based cropping patterns

On all India basis thirty rice based cropping patterns has been identified in different states. Rice is grown in sequence with cotton, pulses, gingelly, jute, wheat, sugarcane, banana, turmeric, betelvine etc.,

2. Millets based cropping patterns

(2a) Maize based cropping patterns: Maize is grown in sequence with sugarcane, groundnut, cowpea, cotton etc. In India twelve cropping patterns were identified with maize.

(2b) Sorghum based cropping pattern: Seventeen patterns were in practice in India.

(2c) Pearlmillet based cropping patterns: Twenty cropping patterns were identified.

Both sorghum and pearlmillets are grown mostly under identical environmental conditions. Sorghum and pearl millet are grown with redgram, groundnut, etc.,

3. Cotton based cropping patterns

On All India basis sixteen cropping patterns are identified with cotton. It is grown with groundnut, sorghum, minor millets, rice, sugar cane and tobacco.

B) Rabi season cropping patterns

Among *rabi* crops wheat, barley, oats, bengal gram, sorghum are the main base crops. Generally wheat and gram are concentrated in the sub-tropical region in the North India, where as rabi sorghum is grown mostly in deccan.

i) Wheat and gram based cropping system: On all India basis ninteteen cropping patterns were identified with wheat and seven with gram. Wheat and gram are grown in sequence with maize, rice, sorghum, millets, groundnut, pearlmillet, cotton etc.

ii) Rabi sorghum based cropping system: Thirteen cropping patterns were identified with rabi sorghum which are grown in sequence with pulses, oilseeds, rice, tobacco, groundnut etc.

(For more details refer annexure IV)

Cropping Pattern in Tamil Nadu

a) Wet land cropping patterns

1.	Rice (June-september)	Rice (October-February)	Rice (February-July)
2.	Rice (July-Oct)	Rice (October-February)	Cotton (February-July)
3.	Rice	Rice	Blackgram/Greengram/ Gingelly
4.	Rice	Rice	Groundnut

b) Garden land cropping patterns

1.	Cotton (Aug-Feb)	Sorghum (February-March)	Ragi (June-August)
2.	Sorghum (Aug-Nov)	Fodder sorghum (Nov-Jan)	Cotton (February - July)

c) High intensive cropping system for Garden lands

It can be achieved by inclusion of an intercrop in each component crop of the cropping pattern.

Sorghum + Cowpea (February – May) – **Onion + Ragi** (June – August) - **Cotton + Onion / Blackgram** (August – February)

d) Dryland cropping patterns

1. **Sorghum + Cowpea / Blackgram**
   (South West Monsoon)
2. Sorghum + Redgram
   (SWM)
3. Sorghum – ratoon sorghum / fodder sorghum
   (SWM) (NEM)
4. Sorghum –Horsegram / Lablab
   (SWM) (NEM)
5. Groundnut + Redgram
   (SWM)
6. Finger millet + Lablab
   (SWM)
7. Cotton + Blackgram / Greengram
   (SWM)

e) Multi-tier cropping in drylands

Three tier cropping of castor, redgram and groundnut and castor-cotton-coriander or blackgram mixtures is popular for a number of years.

Questions

Fill the blanks

1. Classification of crops based on Based on Life cycle is called ____________
2. Raising a crop with the regrowth of the harvested crop is called as ________________
3. The yearly sequence and spatial arrangement of crops or of crops and fallow on a given area is ______________.
4. Growing two or more crops on the same field in a year is called __________.
5. Main and subsidiary crops are sown in rows in ______________

Choose the correct answer

6. Growing two or more crops with definite row proportion at the same time is called as

   a. Mixed cropping b. Intercropping

   c. Relay cropping d. Sequential cropping

7. Growing two or more crops in a piece of land one after another in a year is called as

   a. Multiple cropping b. Sequential cropping

   c. Intercropping d. Relay cropping

8. The second most important pulse crop of India is

   a. Bengal gram b. Pigeonpea

   c. Blackgram d. Greengram

9. Crop plants which on growing exhaust the field because aggressive nature are

   a. Restorative crops b. Residual crops

   c. Catch crops d. Exhaustive crops

10. Crop plants that complete the life cycle within a season or year are called

    a. Annual crops b. Biennial crops

    c. Perennial crops d. None

Answer the following

1. Millets
2. *Rabi* season crops
3. Smother crops
4. Trap crops
5. Multiple cropping.

5

Factors Affecting Crop Production

Factors Affecting Crop production: are classified under the following types,

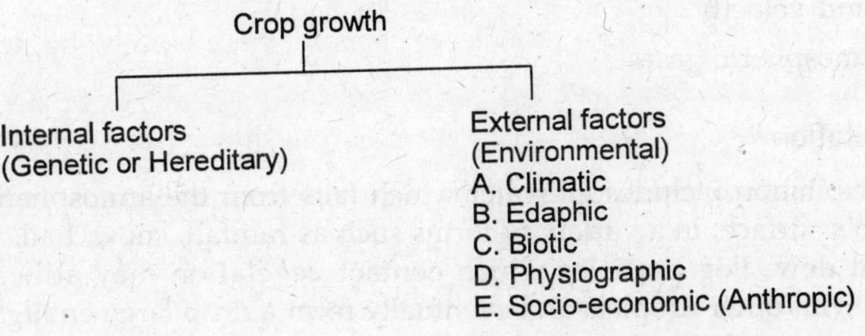

Internal Factors

The increased yield and other desirable characters are related to the genetic make up of the plant. They are :

- High yields under given environmental conditions.
- Early maturity (in some cases late maturity).
- Resistance to lodging.
- Drought, flood and salinity tolerance.
- Tolerance to insects and diseases.
- Chemical composition of grains (high percentage of oil, increase in protein quantity or quality, etc.)
- Quality of grains (fineness, coarseness, etc.)
- Quality of straw (sweetness, juiciness, etc.)

External (Environmental Factors)

Life of crop is so intimately related with the environmental factors of a place.

Environmental factors do not act in isolation from one another. All these environmental factors as discussed below interact with one another to influence the crop growth and production.

A. Climatic factors

The atmospheric factors which affect the crop plants are called climatic factors. They are:

i) Precipitation
ii) Temperature
iii) Atmospheric humidity
iv) Solar radiation
v) Wind velocity
vi) Atmospheric gases

i) Precipitation

Precipitation includes all water which falls from the atmosphere to the earth's surface, in a variety of forms such as rainfall, snow, hail, fog, drip and dew. Fog particles which contact vegetation may adhered, coalesce with other droplets and eventually form a drop large enough to fall to the ground. Condensation of the water vapour present in the air in cool nights results in a deposit of dew.

Rainfall is one the most important factors influencing the vegetation of a place. Most of the crops receive their water supply from rain water. Rain water is the source of soil moisture so essential for the life of a plant. The yearly precipitation, both in total amount and seasonal distribution greatly affects the choice of cultivated crops of a place. Low and ill-distributed rainfall are common features of dry land farming wherein drought resistant crops like sorghum, pearl millet, italian millet and other minor millets are commonly grown. On the other hand in places of heavy and regular rainfall areas such as in the Western Ghats of India crops like rice are grown in flat areas while tea, coffee, rubber etc. are grown in the slopes.

The interaction of rainfall with temperature has a very profound effect on the vegetation and soil of a place. Very heavy rainfall and high temperature in the equatorial and tropical zones, cause the formation of most highly developed vegetation of the world.

Relationship between Climate, Natural vegetation and Soil

Dry cold			Wet cold
Perpetual snow and ice	-	-	Perpetual snow and ice
Tundra	-	Lichens and low shrubs	Tundra soils
Taiga	-	Coniferous forest	Podzol soils
Arid	Semi arid	Subhumid	Humid
Desert grasses and shrubs	Steppe	Grasslands	Rain forests
Sierozems and desert soils	Chestnut and brown soils	Prairi soils and chemozems	Podzol, grey brown/red podzol soils Laterite soil
Dry hot			Wet hot

It may thus be seen that while desert grasses and shrubs are found in desert soils of dry hot climate, rain forests are seen in laterite soils of wet hot climate.

ii) Temperature

Temperature is a measure of intensity of heat energy.

The range of temperature for maximum growth of most of the agricultural plants in between 15 and 40^{0}C.

The temperature of a place is largely determined by its distance from the equator (latitude) and the altitude. Based on the above, the vegetations are classified as tropical, temperate, taiga, tundra and polar.

Some investigators have classified the vegetative of the world into four classes on the basis of prevailing temperature conditions as shown in the following table.

Classification of vegetation based on temperature

Class	Region	Temperature conditions	Type of vegetation	Common crops cultivated
Megatherms	Equatorial and tropical	High temperature throughout the year	Tropical rain	Tropical crops like cassava, varieties of rubber, rice, etc.
Mesotherms	Tropical and sub tropical	High temperature alternating with low temperature of winter	Tropical deciduous forests	Sub-tropical crops like maize, sorghum etc.
Microtherms	Temperate and altitude plants (upto 12,000 ft of tropical and sub-tropical)	Low temperature	Mixed coniferous forests	Temperate crops like wheat, oats, potato
Hekistotherms	Arctic and Alpine regions (above 16,000 ft in tropics and 12,000 ft in temperature regions)	Very low temperature	Alpine vegetation	Pines, Spruce, etc.

Every plant community has its own minimum, optimum and maximum temperature known as their **cardinal points**. Apart from the reduction in yield many injuries on the plants and adverse effect on soil conditions occur under both extremes on temperature.

Cardinal temperature of certain crops for germination

Crops	Minimum °C	Optimum °C	Maximum °C
Wheat	4.5	20	30 – 32
Barley	4.5	20	29 – 30
Oats	4.5	20	29 – 30
Maize	8 – 10	20	40 – 43
Sorghum	12 – 13	25	40
Rice	10 – 12	32	36 – 38
Tobacco	12 – 14	29	35

Cold Injury

Plants are injured due to very low temperature in the following ways:

1. **Chilling injury**: Plants growing in hot climate, if exposed to low temperature (which is above the freezing point) for some time, are found to be killed or injured severely. Example – Chlorotic condition or bands on leaves of sugarcane, sorghum and maize when exposed for 60 hours at 2 to 4^{0}C.

2. **Freezing injury**: This is generally caused in plants growing in temperate region. When the plants are exposed to very low temperature, water is frozen into ice crystals in the intercellular spaces. With further fall in temperature, water is withdrawn from the cells, resulting in increase in the size of ice crystals in the intercellular spaces. The protoplasm of the cell is dehydrated, and mechanical distortion takes place resulting in killing of the cells. Frost damage to potatoes, tea etc. in winter in the hilly areas like The Nilgiris is a typical example of the freezing injury.

3. **Suffocation**: During winter the ice or the snow form a thick cover over the ground and the crop suffers for want of oxygen. Ice in contact with roots inhibits diffusion of carbon dioxide and the respiratory products may become harmful to plants.

4. **Heaving**: Injury to plants is caused by a lifting upward of the plant along with the soil from its normal position in temperate regions where snowfall is common.

Heat Injury

Very high temperature often stops growth. The plant faces incipient starvation due to high respiration rates. The plant is stunted and if such a condition persists for a long period the plant is killed. Direct temperature effects are noticeable in young seedlings and transplanted crops. High temperature causes sterility in flowers. The general effects of excessive heat are defoliation, premature dropping of fruits and in extreme cases death of plants.

Effects of Temperature on Crop Production

Plants can grow only within certain limits of temperature. For each species and variety there are not only optimal temperature limits, but also optimal temperatures for different growth stages and functions, as well as lower and upper lethal limits.

During photosynthesis there are certain biochemical processes preceding and following the reduction of carbon dioxide, which are affected mainly by temperature. As long as light is limiting, temperature has little effect on the rate of photosynthesis. When light is not limiting, it has a profound effect on the rate of photosynethesis.

In general, high temperature accelerates growth process. Rarely are high temperatures *per se* the direct cause of death of plants, provided the water supply is adequate.

Retardation of growth and difficulties in fertilization are observed even in heat loving crops such as sorghum, at extremes of temperatures. The harmful effects of excessive temperatures are usually aggravated by lack of available moisture. Hot dry winds will further increase the damage. Increasing temperatures increases evapotranspiration. The daily alternations of high and low temperature may effect grain production.

(iii) Atmospheric humidity

Water is always present in the atmosphere in the form of invisible water vapour, normally known as humidity of the air.

When the atmosphere contains the maximum possible amount of water vapour it is said to be saturated at the particular temperature and pressure. Any increase in temperature, water remaining constant, will make the air unsaturated. In unsaturated condition, the water vapour content of air is usually expressed as *relative humidity* which is the ratio between the actual humidity present and the saturation humidity possible at that temperature.

The relative humidity of a place, is being affected by temperature and pressure. It is also affected by wind, exposure to radiation, vegetation and water content of the soil. The evaporation of water from plants or a body of water is directly dependent on the relative humidity of the atmosphere. Evapotranspiration of crop plants increases with temperature but decreases with high relative humidity affecting the quantity of irrigation water. Moist air favours the growth of many fungi and bacteria and these affect seriously the crop. The blight diseases of potato and tea are common examples of diseases spread under moist weather. Similarly many kinds of insect parasites such as aphids and jassids thrive well in moist conditions.

(iv) Solar radiation

Solar energy provides two essential needs of plants (a) **light,** required for photosynthesis and for many other functions of the plant-including seed germination, leaf expansion, growth of stem and shoot, flowering, fruiting and even dormancy and (b) **thermal conditions** required for the normal physiological functions of the plant.

Light is one of the most important climatic factors for many vital processes of the plant. It is indispensable for the synthesis of the most important pigment of the plant, i.e. the green chlorophyll. The chlorophyll is capable of absorbing radiant energy and converting it into potential chemical energy of carbohydrates. The carbohydrates manufactured by the plants during photosynthesis are the only link between the solar energy and the living world. It regulates the rate of transpiration by controlling the opening and closing of stomata.

Light affects the plants through its intensity, quality (wave length), duration (photoperiod) and direction.

Light intensity: The variations in light intensity are always accompanied with change in temperature and relative humidity and therefore it is difficult to evaluate light effects alone. Generally speaking light intensity falling at a particular place is normally enough for the plants and their physiological phenomena viz., photosynthesis.

In photosynthesis about one percent of the light energy is converted into potential chemical energy. Very low light intensity reduces the rate of photosynthesis and may even result in the closing of the stomata. This result in reduced vegetative growth of the plants.Very high light intensities is detrimental to plants in many ways. It increases the rate of respiration and thus disturbs the photosynthesis - respiration balance. It causes rapid loss of water resulting in the closure of stomata. The most harmful effect of high light intensity is the phenomenon of "Solarization" in which all the cell contents are oxidized by atmospheric oxygen. This oxidation is different from respiration and is termed as **photo oxidation.**

Depending upon the nature of the crops such as **heliophytes** (sun-loving) and **sciophytes** (shade loving), the dry matter production is affected. Many species produce maximum dry matter under high light intensity if water is available in plenty. However crops like betelvine, turmeric, buckwheat, and tobacco grown during summer produce greater dry matter if slightly shaded.

Quality of light: When white light is passed through a prism it is dispersed into wavelengths of different colors; violet 400-435 millimicrons (mμ); blue 435-490; green 490-574; yellow 574-595; orange 595-626 and red 626 to 750 mμ. The principle wavelengths absorbed in photosynthesis are in the violet-blue and the orange red regions. Among these short rays beyond the violet such as Ultra Violet rays, X rays and Gamma rays and longer rays beyond red such as infrared are detrimental to growth. Red light seems to be the most favourable light for growth followed by violet-blue.

Ultraviolet and shorter wavelengths of the visible light are scattered and infrared and longer wavelengths are absorbed by moisture of the atmosphere. Such a light is called diffused light or sky light. Ultraviolet and shorter wavelengths kill bacteria and many fungi.

Duration of light: It is important from the farmer's point of view in selecting the variety of a crop. The length of the day has greater influence than the intensity. The response of plants to the relative length of day and night is known as '**photo periodism**'. Plants which develop and produce normally when the photoperiod is greater than a critical minimum (more than 12 hours of illumination) are called '*long day plants*' (sugar beet, wheat, barley) and those develop normally when the photoperiod is less than a critical maximum (less than 12 hours of illumination) are called '*short day plants*' (soybeans, maize and millets) Some plants are found to be unaffected by photoperiod and are called as '*day-neutral*' plants (Tomato, asparagus etc.)

Plant characters like floral development, floral initiation, bulb formation, rhizome production etc. are all influenced by photo-periodism. If a long day plant is subjected to short day periods the internodes may be shortened to give a rosette appearance and flowering will not take place. In the same way when a short day plant is subjected to long day period, the growth of the plant becomes abnormal and there is no floral initiation (For example CO 36 rice variety).

Direction of light: Shoots, roots and leaves show different orientation to the direction of light. In temperate regions, the southern slopes show better growth of crops than the northern slopes due to the direction of light contributing more sunlight towards the southern side.

(v) Wind velocity and its effect on crop production

The velocity of wind at a place depends on various factors such as geographical situation, topography, altitude, distance from sea shore, flat plains, vegetation etc. Wind affects crop growth mechanically (directly) and physiologically (indirect).

Mechanical effects

i) Wind causes mechanical lacerations and bruises on the tissues of crop plants,

ii) Violent winds causes lodging of crop plants such as wheat, maize, sugarcane, paddy etc.,

iii) A very high velocity of wind (gale, blizzard, hurricane, cyclone etc.) breaks dead and living branches of trees and even uproots them completely,

iv) In bare deserts, high velocity of wind causes constant soil erosion and this makes it difficult for plants to grow,

v) Wind has a powerful effect on the humidity of atmosphere.

Physiological effects

i) Wind increases the rate of transpiration in plants,

ii) Hot dry winds causes much damage to crops at the time of flowering

iii) The internal water balance of plants is affected resulting in poor seed setting.

iv) Another form of damage in *blossom injury* caused by evaporation of secretions from the stigmas.

Beneficial effects

i) Wind is also responsible for causing rainfall to a very large extent. In India the monsoon type of rainfall is largely determined by particular patterns of wind movement. (Trade winds)

ii) Wind helps in pollination of flowers and dispersal of seed, fruits and microorganisms.

iii) The hot, dry wind may reduce the incidence of dangerous yellow rust of wheat,

iv) Moderate winds have a beneficial effect on photosynthesis by continuously replacing the carbon dioxide absorbed by the leaf surfaces.

Remedy

i) Velocity of wind can be reduced by growing tall, sturdy trees across the direction of the wind as *wind breaks*

ii) To arrest the movement of soil by wind erosion *shelter belts* of vegetation are raised.

vi) Atmospheric gases

The atmosphere surrounding the earth contains a mixture of gases viz., carbon dioxide (0.03%), oxygen (20.95%), nitrogen (78.09%), argon (0.93) and miscellaneous gases (0.02%) in a constant proportion.

Carbon dioxide: The CO_2 is the main source of carbon for the various types of organic compounds in the body of the plants. It is the main raw material for the manufacture of carbohydrates by the photosynthetic process of the green plants. Photosynthesis is approximately proportional to the concentration of CO_2 in the air surrounding the foliage of the crop. The CO_2 which gets incorporated into the organic compounds of the plants returns to the atmosphere by respiratory break down of these compounds and by the death, decay and combustion of plants (carbon cycle).

Increased growth and greater yield of vegetables are possible under green house conditions by increasing the content. Aquatic plants utilize the dissolved CO_2, from water.

Oxygen: Life sustains because of oxygen. The amount of O_2, is normally constant in the air because plants give off O_2, during photosynthesis.

Nitrogen: Lightning, rainfall and nitrogen fixing microorganism contributes nitrogen to the soil from the atmosphere. Symbiotic bacteria like rhizobium, free living bacteria like azotobacter, blue green algae etc. fix a good amount of nitrogen in the soil. The decomposition of dead plants and animals also adds N to the soil. The N in the soil is made available to the crops by the activity of nitrifying bacteria.

Certain gases like SO_2, CO and HF when released into the air in sufficient quantities are toxic to plants.

B. Edaphic factors

Plants grown in a land are completely dependent on the soil in which they grow for anchorage, water and mineral nutrients.

The soil factors which affect the crop growth are,

1. Soil moisture
2. Soil air
3. Soil temperature
4. Soil mineral matter
5. Soil organic matter

Composition of soil	
Composition	**Percent**
Mineral matter	30
Soil moisture	30
Soil air	30
Soil organic matter	5-10

6. Soil organisms and
7. Soil reaction.

1. Soil moisture

In plant tissues water constitute about 90%. The moisture lost through transpiration is made up only by absorbing water from the soil. The moisture is held within the soil by the attractive forces of the soil particles. The capillary moisture held in the pF range of 2.54 to 4.2 is available for plants. It is the available soil moisture range between saturation point and wilting point. The moisture with about pF 2.54 is very favourable for plant growth. The moisture near pF 4.2 (wilting point) is absorbed by plants only with great difficulty and the plants may not be able to make vigorous growth near this moisture level.

(pF: is the logarithm of height (in cm) of a water column that represents total stress with which water is held by the soil)

2. Soil air

Aeration of the soil is absolutely essential for the absorption of water by roots. O_2 is required for respiration of roots and microorganisms. In poorly aerated soil the CO_2 get accumulated and is detrimental, for absorption of water by the roots. Soil air is also useful in increasing the nutrient availability of the soil by,

i) breaking down the insoluble mineral into soluble salts;
ii) decomposing organic matter and
iii) bringing out nitrifying (nitrogen releasing) and nitrogen fixing processes of bacteria.

3. Soil temperature

1. It affects the physical and chemical processes going on in the soil.
2. It influences to a very great extent
   a) the rate of absorption of water and solutes
   b) the germination of seeds and
   c) the rate of growth of the underground portions of the plant body.

The maximum absorption of water by the roots take place generally between 20^0 and 30^0C. Temperature below 20^0 C causes appreciable reduction in the rate of absorption of water. Cold soils are therefore not conducive for rapid growth of most agricultural crops.

3. Soil temperature controls microbiological activity and processes involved in the availability of nutrients to plants. Nitrification begins in the soil when the temperature reaches about 5^0C.

4. Soil mineral matter

The mineral content of the soil is derived from the weathering of the rocks and minerals as particles of different sizes. These are sources of plant nutrients such as Si, Ca, Mg, Fe, K, Na and Al. Minor elements (trace elements) like B, Mo, Zn, Cu, Co, I and F are also present in very small quantities.

5. Soil organic matter

The soil organic matter content varies from less than 1% in arid sandy soils to as much as 90% of dry weight of well developed soil. It has a marked influence on the soil properties and growth of crops. The organic matter of the soil is derived from i) dead and decaying roots of plants and living organisms present in the soil and ii) dry leaves, twigs, dead plants and animals added to the soils.

Advantages of soil organic matter

a) It is the source of essential plant nutrients for crop growth. It contains 95% of total N, 50-60% of total P and 10-20% of the total S;

b) It increases water holding capacity of the soil due to its organic colloids;

c) It increase the cation exchange capacity (CEC) of the soil; and

d) It is a source of food for most of the soil organisms.

6. Soil organisms

Raw organic matter in the soil is not directly used by the plants as food. It must be broken down first into humus and then into simpler products before it can be utilized. This work is done by different kinds of organisms as given below which inhabit the soil in billions.

Flora	**Macro flora-** Roots of higher plants
Micro flora	Bacteria, Actinomycetes, Fungi, Algae

Fauna-Macro fauna	Earthworms,Burrowing vertebrates (moles, gophers, rats, etc.)
Micro fauna	Protozoa, Nematodes, Mites, Insects,

7. Soil reaction

Soils may be neutral, acidic or alkaline depending upon their content of basic salts and acidic components. Neutral soils are the best for growth of most crops. Soil acidity beyond a particular limit **(soil with low pH)** is injurious to the plant growth due to

a) Aluminium toxicity under high acidity

b) Interferes with the absorption of several nutrients particularly cations like K, Ca and Mg.

Effects of high soil pH

a) P gets fixed in acidic soils.

b) Organic matter decomposition is reduced.

c) Activities of nitrifying and nitrogen fixing bacteria may be checked.

d) Favours fungal diseases like potato scab.

Similarly high alkalinity of the soil(soil with high pH) also adversely affects crop growth. Presence of high amount of sodium interferes with nutrient and water absorption. Soil aeration and infiltration are also reduced due to poor physical condition.

C. Biotic factors

Beneficial or harmful effects caused by other plants and animals on the crop plants are the effect of biotic factors.

1. Plants

a) *Competitive and complimentary nature among field crops:* Competition between plants occurs when there is demand for nutrients, moisture and sunlight, particularly when they are in short supply or the plants are closely spaced. Optimum spacing of crops is an important agronomical practice. When different crops such as cereals and legumes are grown together as in mixed cropping there is mutual benefit resulting in better yield.

b) *Competition between weed and crop:* Weeds reduce crop yields due to competition with crops for water, soil nutrients and light. In dry farming condition weeds compete with crops for water. In irrigated tracts, the competition is severe for nutrients. Weeds in fallow land deplete the soil of both moisture and nutrients.

c) *Plants as Parasites:* A plant parasite is dependent on its host plant for its existence. Parasitic plants like striga, orobanche, cuscuta and loranthus live on the host plants and affects growth of the cop plants. Parasitic fungi, bacteria, virus etc. causes different kinds of diseases on agricultural crops.

d) *Symbiosis:* Different organisms have mutual relationship with each other and with the environment. This biological inter relationship among the organisms is termed as symbiosis. Example – Legumes and rhizobia – nodule forming; Azoto bacter – free living (it fixes elemental N present in the atmosphere and supplies to the plants).

2. Animals

a) Soil animals or Soil fauna

Include protozoa, nematode, rotifiers, snails and insects. They help organic matter decomposition while using the organic matter for their living.

i) **Harmful organisms:** Insects and nematodes cause considerable damage as crop pests during the growth of plants and in storage of grains. The average loss due to insects is about 20% throughout the world.

ii) **Beneficial organisms:** Many plants are pollinated primarily by insects. The examples of beneficial organisms are,

1. Bees and wasps are important for pollination.
2. Moths, butterfly and beetles also do pollination. Beetle (*Elaeis kaemeniricus*) is necessary to have good pollination in oil palm crop.
3. Burrowing of earthworm facilitate aeration and drainage of the soil, Ingestion of organic matter and mineral matter results in a constant mixing of these materials in the soil there by favours better growth of plants.

b) Small animals

Small animals like rabbits, squirrels and field rats also cause excessive damage to field and garden crops.

c) Large animals

Large animals like domestic animals and wild animals cause damage to crop plants by grazing and browsing habits.

D. Physiographic Factors

It can be studied under two catagories such as :

1. **Geological Strata:** It accounts not only for the kind of parent material utilized in soil formation but also on the nature of crops grown in these soils for proper utilization.

2. **Topography:** The nature of the surface of earth is known as topography. Topographic factors affect the crops indirectly by modifying climatic and edaphic factors of a place. It includes,

   a. *Altitude of the place:* Increase in altitude causes decrease in temperature and increase in precipitation and wind velocity.

   b. *Steepness of slope:* It causes swift run off of rain water resulting decreased moisture content of soil. The organic matter of the soil increases resulting in high N content and acidity of the soil.

   c. *Exposure of the slope to light and wind* – A mountain slope exposed to weak intensity of light and strong dry winds as the case of northern slopes of temperate regions and the himalayas may have poor crops due to want of moisture and sunlight. Similarly the western slopes of Tamil Nadu hills poor crops due to damage caused by heavy winds.

   d. *Direction of the mountain chains:* It governs the distribution of the rainfall during monsoon and also the type of crops in dry farming.

E. Anthropic (socio – economic) Factors

1. Man/women produce changes in plant environment and are responsible for scientific crop and soil management,
2. breeding varieties for increased yield and
3. introduction of exotic plants

These factors affect the management of soil and crop which leads to higher production. In addition to the above the socio economic factors affecting the crop production are as follows,

i) the economic conditions of the farmer greatly decides the input/ resource mobilizing capacity,

ii) the educational status and technical know-how of the farmer,

iii) the resource allocation ability and social values of the farmer,

iv) government price policy,

v) marketing and storage facilities etc.

Questions

Fill the blanks

1. The increase in crop yields and other desirable characters are related to ______________of plants.
2. ______________ is the most important form of precipitation which influences the vegetation of place
3. The range of temperature for maximum growth of most of the agricultural plants is between ______________
4. ______________ rays are detrimental for crop growth
5. Every plant community has its own minimum, maximum and optimum temperature known as __________________

Choose the correct answer

6. Response of crops to relative length of day and night is called

   a. Phototropism b. Photorespiration

   c. Photoperiodism d. Geotropism

7. Climatic factors affecting crop production are otherwise called as

   a. Genetic b. Atmospheric

   c. Environmental d. Hereditary

8. The most favourable light for crop growth is

   a. Red light b. Infra red

   c. Ultra violet d. Yellow

9. Freezing injury is generally caused in plants growing in

   a. Temperate region b. Tropical region

c. Arid region
d. Sub tropical region

10. Lifting upward of the plant along with the soil from its normal position in temperate regions where snowfall is common

a. Suffocation
b. Chilling injury
c. Heaving
d. Freezing injury

Answer the following

1. Heaving
2. Long day plants
3. Day neutral plants
4. Soil organisms
5. Soil air.

6

Soils

Soil is defined as the thin layer of earth's crust made up of disintegrated and decomposed rocks, complex mineral compound, organic matter, water/air and living organism like bacteria, fungi, insects and worms and serves as the natural medium of growth of plants.

- It provides nutrients, moisture, anchorage (support) and provides air to root system.
- There are different soil groups found in varied regions of India.
- Each group differs from other in physical and chemical properties.
- The variation in behaviour is mainly due to the nature of the parent material from which the soils are formed.
- Parent materials are Igneous rocks, sedimentary rocks and metamorphic rocks.
- Physical properties like structure, texture, colour, water holding capacity, depth etc. are to be noted. Chemical properties like the presence of various plant elements, pH, EC, CEC, acidic or alkaline, etc. are considered.

For a farmer soil refers to the cultivated top layer only, that is, upto 15 to 18 cm of the plough depth. It is also referred as surface soil. The layer below the surface is known as subsoil. Due to physical and chemical weathering of surface soil, it has been broken down to a fine condition. But the subsoil is protected by the surface layer and hence there is no weathering. Hence this layer is coarser.

Soils widely vary in their characteristics and properties. Understanding the properties of soils is important (1) for optimum use they can be put to and (2) for best management requirements for their efficient and productive use. In order to establish to interrelationship between their characteristics, they require to be classified. The major soil groups of India are as follows:

Classification Based on Soil Taxonomy

Order	*Entisols*	Suborder	Fluvents
Great Group	Torrifluvents	Subgroup	Typic Torrifluvents
Family	Fine-loamy, mixed, superactive, calcareous		
Series	Jocity, Youngston.		

Major Soils of India

1. Alluvial soil (*Entisols, Inceptisols and Alfisol)*
2. Black soil (*Vertisol*)
3. Red soil (*Alfisol*)
4. Lateritic soil (*Ultisol*)
5. Desert soil (*Aridisol*)
6. Peaty and Organic soil
7. Problem soils (saline, alkali, acid)

1. Alluvial soil or Indo-Gangetic Alluvium

- This is the most extensive soil found in India.
- Out of total area of India, 48.0 m.ha comes under river alluvium.
- These soils include deltaic alluvium, calcareous alluvium and coastal alluvium.
- Alluvial soils are formed by transportation in streams and rivers and are deposited in flood plains or along the coastal belts.
- Newly formed alluvium may not have distinct soil horizons while older alluvium may have soil horizons.
- They occur in the basins of Indus, Ganges, Brahmaputra, Godavari, Krishna, Cauvery and Tamiraparani deltas spread in U.P., Bihar, West Bengal, Gujarat, Punjab, Rajasthan, Andhra Predesh, Tamil Nadu.
- Newer alluvium is called as *Khadar*, is sandy, light colour and less *Kankar* nodules.
- Older alluvium is called as *Bhangar*, full of clay, dark colour and more *Kankar* nodules.

- Alluvial soils of high altitude are acidic in nature and plains are neutral to alkaline.
- Alluvial soils of plains are medium in phosphorous content and high in potassium content. Generally, alluvial soils are rich in nutrients and are fertile and they support good crop growth with plenty of water.
- Many crops including vegetable are cultivated in river alluvium.
- Crops like rice, wheat, cotton, maize, sugarcane, vegetables, jute, oil seeds, millets, pulses and fruits are cultivated in these soil.

2. Black soil

- Dark-grey in colour due to clay-humus complex.
- Area of around 32.0 m.ha is under this soil.
- This soil is also called black cotton soil, mixing of soil along the entire column with *Montmorillonite* clay.
- Cotton grows very well with water available in soil.
- Black soil holds more moisture and available for a long time.
- Found in Maharashtra, Madhya Pradesh, South Orissa, South and Coastal Andhra Pradesh, North Karnataka and parts of Tamil Nadu.
- Black soil contains high proportion of clay (30-40%), so, the water holding capacity is high.
- Typical characteristics of this black soil are swelling (during wet period) and shrinkage (dry period).
- While dry, it forms very deep cracks of more than 30-45 cm.
- In Kovilpatti (Tamil Nadu) areas the cracks may extend to 2 to 3 m with a width of 1 to 6 cm.
- Field preparation takes longer time compared to other soil.
- Only after secondary tillage, the soil is suited for crop production.
- The soils are fine grained contain high proportion of Calcium and Magnesium carbonates.
- They are poor in N, medium in P and medium to high in K (Characteristic feature of typical Indian soil).
- In Tamil Nadu black soils have high pH (8.5 to 9) and are rich in lime (5-7%), have low permeability.

- The soils are with more cation exchange capacity (40-60 m.e./ 100 g).
- Crops grown in this soil are cotton, bengal gram, mustard, millets, pulses, oil seeds (sunflower, safflower) are commonly grown in this soil.
- Most of the soils come under rainfed areas.

3. Red soil

- Based on the colour (due to presence of ferric oxides) it is called as red soil.
- Around 30 m.ha are found in India. They are formed from granites and other metamorphic rocks.
- Mostly found in semi-arid areas and the colour varies from red to yellow.
- The soil is light textured, with *Kaolinite* type of clay. Well drained with moderate permeability.
- Low cation exchange capacity and low water holding capacity.
- Red soil is present in Gujarat, Tamil Nadu, Karnataka, Andhra Pradesh, North and East of Arunachal Pradesh, Madhya Pradesh, Parts of Bihar and Uttar Pradesh.
- They are shallow in depth because they are degraded or drained soil.
- Lesser clay and more sandy than *Vertisol*.
- Red soil is always in acidic nature.
- Highly suitable for groundnut crop cultivation.
- Crops like millets, pulses, oil seeds (ground nut, gingelly, castor) and tuber crops like cassava are commonly cultivated.

4. Laterites and Lateritic soil

- Laterite soils are formed due to the process of laterisation. i.e., leaching of all cations leaving Fe and Al oxides.
- Mostly found in hills and foothill areas.
- This soil is formed under high intensive down pour of rainfall.

- It is modified form of red soil, clay content is minimum.
- Rich in organic matter content and rich in fertility and medium water holding capacity.
- They become very hard when there is no water.
- The cohesive nature is high.
- Acid loving crops (Plantation crops) and fruits (pineapple, avacado) are more cultivated. Tea, rubber, pepper, spices are cultivated.
- At lower elevations, rice is grown.

5. Desert soil

- Found in desert regions of Rajasthan (Thar desert), parts of Haryana and Punjab of India.
- More sand is found and sand dunes are common.
- Clay content is < 8% only.
- Poor fertility, poor water holding capacity and susceptible to soil erosion.
- Presence of sodic salts (high Na content) leads to alkalinity.
- Crops like date palm, cucumber, millets are cultivated (countries like Saudi Arabia, UAE, Jordan, Sudan etc).

6. Peaty and Organic soil

- These soils are very rich in organic matter.
- Found in Kerala, coastal regions of West Bengal, Orissa, South and East coast of Tamil Nadu. Deposition of organic matter by the elevated soil.
- Peaty and organic soil is not suitable for majority of crops.
- Rice is mostly cultivated in coastal area in rainy season.

7. Problem soil

a. Saline soils

- Contain excess amounts of neutral soluble salts dominated by chlorides and sulphates of Na, Ca and Mg affects plant growth.
- White encrustation of salts is formed and hence called white alkali.

- These soils are characterized by, EC: $4dSm^{-1}$ at 25°C, ESP: < 15; pH; < 8.5.
- This soil needs leaching and drainage before cropping for amelioration.

High salt tolerant: *Sesbania*, Rice, sugarcane, oats, berseem, lucerne, indian clover & barley.

i) Medium salt tolerant: Castor, cotton, sorghum, pearl millet, maize, mustard & wheat.

ii) Low salt tolerant: Pulses, peas, *Sunnhemp*, gram, linseed and sesame.

b. Sodic / Alkali soils

- High content of carbonates and bicarbonates of Na.
- Hence, they are with high exchangeable sodium percentage (ESP) with dark encrustation, hence called as black alkali.
- These soils are rich in $NaHCO_3$ and characterized by pH: > 8.5; EC: < $4dSm^{-1}$; ESP : > 15.
- Use gypsum ($CaSO_4$, $2H_2O$) as amendment for reclamation of sodic alkali soils. Iron pyrites (FeS_2), bulky organic manures (especially green manure) and crop residues which produces weak organic acids.

i) Tolerant crops: Karnal/rhodes/para/bermuda grass, rice and sugar beet.

ii) Semi-tolerant: Wheat, barley, oats, berseem and sugarcane.

iii) Sensitive: Cowpea, gram, groundnut, lentil, peas and maize.

c. Acid soils

- These are low pH with high amounts of exchangeable H^+ and Al_3.
- Occur in regions with high rainfall.
- Significant amount of partly decomposed organic matter exist.
- Have low CEC and high base saturation.
- Liming and judicious use of fertilizers are the management measures suggested.
- Suitable crops: Acedophytes (like potato).

Comparison of three types of soils

Parameters	Saline soil	Saline alkali	Alkali soil
EC (dS/m)	>4	>4	<4
ESP (%)	<15	>15	>15
pH	<8.5	<8.5	>8.5

Major Soils of Tamilnadu

In Tamil Nadu, the major portion is covered by red sandy soil and red loamy soils. Red sandy soils have developed from acidic parent material like granite, gneiss, quartzite, sandstone etc. The red colour of soils is due to the coating of ferric oxides on soil particles. Sand particles are coated with red coloured hematite or yellow coloured limonite which is responsible for the various shades of red and yellow of these soils, which usually contain ferruginous gravel containing iron, aluminium and silica. These sandy, loamy sand and sandy loam soils are heavily leached and therefore poor in basic elements and plant nutrients.Their pH ranges from 6.6 to 8.0. Calcium is the important exchangeable cation in these soils. They are neutral to slightly alkaline in reaction.

Soils of Tamilnadu

Type of Soil	Areas in Tamil Nadu
Red loam (79.8 L. ha & 61.7%)	Parts of Kancheepuram, Cuddalore, Salem, Dharmapuri, Coimbatore, Tiruchirappalli, Thanjavur, Ramanathapuram, Madurai, Tirunelveli, Sivagangai, Thoothukudi, Virudhunagar, Dindigul and The Nilgiris Districts.
Laterite soil (3.8 L.ha & 2.9%)	Parts of The Nilgiris District
Black soil (15.0 L. ha & 11.6%)	Parts of Kancheepuram, Cuddalore, Vellore, Thiruvannamalai, Salem, Dharmapuri, Madurai, Ramanathapuram, Tirunelveli, Sivagangai, Thoothukudi, The Nilgiris, Virudhunagar and Dindigul Districts.
Sandy coastal alluvium (9.8 L. ha & 7.6%)	On the Coasts in the districts of Ramanathapuram, Thanjavur, Nagapattinam, Cuddalore, Kancheepuram and Kanyakumari
River alluvium (21.0 L. ha & 16.2%)	All river deltaic areas (Cauvery, Vaigai, Tamiraparani)

Source: Department of Economics and Statistics, Chennai

Soil classification

Type of Soil	Areas in Tamil Nadu
Red Loam	Erode, Namakkal, Salem, Dindigul and Coimbatore
Laterite Soil	Dharmapuri, Kancheepuram, Thiruvannamalai, Thiruvallur and Vellore
Black Soil	Thoothukkudi, Virdhunagar, Pudukkottai, Madurai, Dindigul, Tiruchirappalli, Ramanathapuram, Karur. Namakkal, Theni and Salem
Sandy Coastal Alluvium	Pudukkottai, Madurai, Dindigul, Perambalur, Thanjavur and Tiruchirappalli
Red Sandy Soil	Coimbatore, Tiruchirappalli, Karur, Theni, Madurai, Kancheepuram, Tiruvannamalai, Thiruvallur and Vellore

Source: Commissioner of Agriculture, Department of Agriculture, Chennai – 600 005.

Questions

Fill the blanks

1. The most extensive soil found in India is ______________
2. The colour of red soil is due to the presence of ___________
3. *Kaolinite* type of clay is dominating in ___________ soil
4. The major soil type in Tamil Nadu is ______________ soil
5. Sodic soils have high content of carbonates bicarbonates of ______

Choose the correct answer

1. Soil type suitable for groundnut crop is

   a. Red soil b. Black soil

   c. Sandy soil d. Saline soil

2. Soil type suitable for cotton crop is

   a. Red soil b. Black soil

   c. Sandy soil d. Saline soil

3. Soil that Contain excess amounts of neutral soluble salts dominated by chlorides and sulphates of Na, Ca and Mg is

   a. Sodic soil b. Alkaline soil

   c. Acid soil d. Saline soil

4. The soil which is rich in organic matter is

   a. Black soil   b. Peat soil

   c. Red soil   d. Sandy soil

5. The reclamation measure for acid soil is

   a. Gypsum   b. Liming

   c. Leaching   d. None

Answer the following

1. Importance of soil
2. Major soils of India
3. Black soil
4. Laterite soil
5. Saline soil.

7

Seasons

Agricultural Seasons

Season is defined as "part of the year during which a distinguished type of weather prevails".

A) Seasons of Temperate region

i) *Spring* (March – May) is the first season of the year in which plants being to grow and leaves emerge.

ii) *Summer* (June – August) is the second and warmest season of the year out side the tropics during which plants flourish

iii) *Fall or Autumn* (September – November) is the third season of the year in which leaves turn brown.

iv) *Winter* (December – February) is the last and the coldest season of the year. Many trees loose their leaves.

B) Seasons of India

According to India meteorological Department (IMD) there are four seasons n a year, in India.

i) ***Kharif*** */ Monsoon or south west monsoon* – June to September

ii) *Post monsoon (north east monsoon)* – October – November and

iii) *Winter – December to February*

iv) ***Zaid*** */ Summer or pre-monsoon* – March to May

The post monsoon (N.E.M) and winter season are combined together and designated as Rabi (October to February) throughout India. Accoridngly the agricultural seasons in India are called as *kharif, rabi and summer.*

C) Seasons of South India

In southern states of India (Tamilnadu, Andhra Pradesh and Karnataka) there is slight variation in the season based on rainfall duration as follows.

i) Cold Weather period (winter)	: January	- February
ii) Hot Weather period (summer	: March	- May
iii) South West monsoon period (Rainy season)	: June	- September
iv) North East monsoon period	: October	- December

Characteristics of Seasons

Winter: The weather prevailing during this period is cool, usually dry and pleasant with dew fall during morning hours. Occasional cyclonic depressions bring light rains to North western regions in India. These rains though in small amount, are most beneficial to the winter crops. At times they do more harm than good, due to shedding of cotton bolls, damaging of tobacco quality.

Hot Weather: This period is characterized by high temperature. The temperatures are higher in the north during this period than South India. Showers during this period are mainly useful for preparatory cultivation (Summer ploughing). Gingelly and to a limited extent sorghum are sown with the rains. The garden land crops are benefited by these rains.

South West Monsoon: This is the rainy season in India except most part of the Tamil Nadu. About 60% of the rainfall in an year is received during this period. Most of the tropical crops (*Kharif* crops) are grown during this period. All dry lands and also wet lands directly depend on the rains received during this period. Garden land crops are also benefited by these rains. The climate prevailing during this period is warm and humid with bright sunshine except on rainy days (Tropical climate).

North East Monsoon: The temperature is high upto the middle of October and later starts falling rapidly.The rainfall received during the period is about 33% of annual rainfall except in Tamil Nadu and Coastal Andhra Pradesh where the annual rainfall received during this period is more than 55% half of it with occasional cyclones. The sky is clear in northern India. Mostly temperate (*Rabi*) crops are grown during this period.

Rice Seasons: Rice is grown in different seasons during a year.

A) Rice seasons in North India: (West Bengal)

i) AUS : May – September

ii) AMAN : June, July – November, December (*Kharif*)

iii) BORO : January – May (Summer)

B) Rice seasons of Tamilnadu

Totally seven seasons which vary with districts ..,

i) In cauvery Delta (Tanjore, Thiruvarur, Nagapattinam,South Arcot (part))

1. **Kuruvai** : June – September
2. **Early Samba** : July/August – December/January

   Samba : August – January

   Late Samba : September/October – January/February
3. **Thaladi** : September – January / February
4. **Navarai** : December – April

ii) In Chengalput and Arcot districts (Tiruvallur, Thiruvannamalai, Vellore, Kanchipuram, Vizhupuram and Cuddalore districts), there is one more season in addition to above four as,

5. **Sornavari** : April/May – August/September

iii) In Tirunelveli and Kanyakumari districts rice is grown in

6. **Kar** : May – September and
7. **Pishanam** : September – January

Cotton Seasons of Tamilnadu

i) *Winter Irrigated* : August – September sowing

ii) *Summer* : February – March sowing

iii) *Rice fallow* : January – February sowing

iv) *Rainfed cotton* : September – October sowing

Sugarcane Seasons of Tamilnadu

a) *Early Season* : December – January planting

b) *Mid Season* : February – March planting

c) *Late Season* : April – May planting

d) *Special Season* : June – July planting

Effect of Season on Choice of Crops

Season influences the crop selection to a greater extent as it decides the growth and establishment of the crop. The weather conditions prevailing during a season totally governs the crop production. The fluctuation in the crop yield depends on the length of the rainy season, the quantity and regularity of distribution of rainfall and the amount of rainfall received after the rainy season. The climatic factors viz., precipitation, wind, solar radiation (light and thermal energy) temperature, atmospheric air and its pressure prevailing in a season very greatly influences the crop growth, establishment and yield as discussed in the earlier chapter on climatic factors affecting the crop production.

Questions

Fill the blanks

1. According to India meteorological Department (IMD) there are ________ seasons in a year, in India.
2. For sugarcane, the special season planting is during __________.
3. The rainy season for majority of the states in india is __________.
4. Rainfall during hot weather period are are mainly useful for _____.
5. In Cauvery Delta zone, Kuruvai season starts in __________.

Answer the following

1. Seasons of India
2. Rice seasons of North India
3. Cotton seasons of Tamil Nadu
4. Sugarcane seasons of Tamil Nadu
5. Rice seasons of Tamil Nadu.

8

Systems of Farming

For better understanding of different systems of farming it is essential to study certain terminologies.

Farm is a piece of land with specific boundaries, where crop and livestock enterprises are taken up under a common management.

Farming is the process of harnessing solar energy in the form of economic plant and animal products or it is the business of cultivating land, raising livestock etc.

System refers to an orderly set of interdependent and interacting components none of which can be modified without causing a related change elsewhere in the system.

There are three district systems of farming as wetland system of farming, garden land (irrigated dry land) system of farming and dryland system of farming.

Wetland: Pertaining to soils flooded or copiously irrigated through lake or pond or tank for a least several weeks in each year (or) to crop growth in such soils. The water is not entirely under the control of the farmer.

Wetland Farming

Wetland Farming is the practice of growing crops in soils flooded through natural flow of water for most part of the year.

Garden land: means the land supplied with water mostly from underground sources i.e. irrigated dryland.

Garden land farming: growing crops with supplemental irrigation by lifting water from underground sources. Crops grown in these lands are irrigated through lift irrigation and hence the water is under control.

Dry Farming / Dry Land Farming: is the practice of crop production entirely with rainfall received during the crop season or with conserved soil moisture and the crop may face mild to severe stress during its life cycle. It is practiced in areas with an annual rainfall less than 800 mm (Arid and Semi arid).

In areas where the rainfall is more than 800 mm (Humid and Sub humid) the crops are grown entirely with the rainfall received during the crop season. The crop may face little or no moisture stress during its life cycle which is known as **Rainfed Farming**.

Low land means the land submerged for most part of the year. Low lands are mostly wet lands either with irrigation (irrigated) or with rainfall (rainfed).

Upland means the land unsubmerged and well aerated. Uplands are mostly dry lands.

Comparision of Some Selected Features of Different Farming Systems

S. No.	Features	Wetlands	Irrigated Land	Dryland	Rainfed
1.	Farming Practices (Duration)	9-12 Months	9-12 Months	<6	6-8 Months
2.	Source of Water	River, Lake Pond/Tank	Wells	Rainfall (800mm/ Year)	Rainfall (>800mm/ Year)
3.	Climate	Arid to Humid	Arid to Humid	Arid to Semi-arid	Subhumid to Humid
4.	Irrigation	Natural Flow	Lift Irrigation	-no Irrigation	
5.	Water Management	Manage-Metn of Excess Water	Economical Water Use	Water Conser Vation	
6.	Fertilizer Management	Liberal Use	Liberal Use	Limited Use	
7.	Objective	Maximising the yield	Yield maximi zation	To get sustainable yield	
8.	Constraints	Soil Health, Salt Affected Soils, Drainage	Salt Affected Soils	Wind & Water Erosion	

The crop production is entirely different in dry-farming and irrigated farming systems due to irrigation as given below.

Difference between dry farming (Rainfed) and irrigated farming

Dry farming	Irrigated farming
1. The field is ploughed deep to increase infiltration of rains	No need of deep ploughing to conserve water.
2. Land is prepared immediately after rainfall	Land is prepared according to optimum time of sowing.
3. Seeds are sown at more depth To make contact with moisture.	Seeds are sown at optimum depth.
4. Crops or crop varieties having drought tolerance or less water requirement are used.	According to the need, crops or their varieties are selected.
5. Generally, short duration crops	Selection of crops depends on the need are preferred.
6. Mixed inter cropping is beneficial.	Generally pure cropping is done.
7. Due to limitation of moisture one or two crops in a year is possible.	More than 2 crops in a year are Grown, subject to the availability of water
8. Crop failure (risk) is expected	No chance of crop failure (no risk)

Sustainable Agriculture

Definition

A farming systems that are "capable of maintaining their productivity and usefulness to society indefinitely and must be resource-conserving, socially supportive, commercially competitive, and environmentally sound."

As per USDA (legal) Definition

Sustainable agriculture means, an integrated system of plant and animal production practices having a site-specific application that will, over the long term :

- Satisfy human food and fiber needs;
- Enhance environmental quality and the natural resource based upon which the agricultural economy depends;

- Make the most efficient use of nonrenewable resources and on-farm resources and integrate, where appropriate, natural biological cycles and controls;
- Sustain the economic viability of farm operations;
- Enhance the quality of life for farmers and society as a whole.

Advantages

- Production cost is low
- Overall risk of the farmer is reduced
- Pollution of water is avoided
- Very little or no pesticide residue is ensured
- Ensures both short and long term profitability

Disadvantages

- Since sustainable agriculture uses least quantum of inputs, naturally the output (yield) may also be less.

Major components of sustainable agricultural system

- Soil and water conservation to prevent degradation of soil productivity
- Efficient use of limited irrigation water without leading to problems of soil salinity, alkalinity and high ground water table
- Crop rotations that mitigate weed, disease and insect problems, increase soil productivity and minimise soil erosion
- Integrated nutrient management that reduces the need for chemical fertilizers improves the soil health and minimise environmental pollution by conjunctive use of organics, in-organics and bio-fertilizers.
- 'Integrated pest management that reduces the need for agrochemicals by crop rotation, weather monitoring, use of resistant cultivar, planting time and biological pest control.
- Management system to control weed by preventive measures, tillage, timely inter cultivation and crop rotation to improve plant health.

Two farming systems have been proposed for enduring sustainability. They are,

I) Low external input sustainable Agriculture or Low input sustainable Agriculture (LEISA / LISA): Minimal use of external production inputs.

Advantages

i) Production costs are low

ii) Overall risk of the farmer is considerably reduced

iii) Pollution of water is avoided.

iv) Healthy food very little or no pesticide residue is ensured.

v) Ensure both short and long term profitability.

Disadvantages

Continuation of LEISA will perpetuate a vicious circle of "low input-low yields" which the third world countries with even increasing population cannot afford. The solution for this is the optimal input farming which will meet the requirement of sustainability wit the promise of low input/unit of output. It lays emphasis on law of diminishing returns.

II) Organic farming

It is a production system which avoids or largely excludes the use of synthetic compound fertilizers, pesticides, growth regulators and live stock feed additives. To the maximum extent feasible, it relies on crop rotation, crop residues, animal manures, legumes, green manures, off-farming organic wastes and aspect of biological pest control to maintain soil productivity and tilth, to supply plant nutrients and to control insects, weeds and other pests.

In this system most of the ill effects of modern day agriculture is avoided because,

i) Use of agrochemical is forbidden.

ii) There is emphasis in building up of organic matter in the soil, there by activate biological activity.

iii) Soil is treated as living organism.

Emphasis is given on

i) Maintenance of favourable soil structure.

ii) Development and use of crop rotation that improve and prevent soil erosion.

iii) Biological control of pests, diseases and weeds.

Need and scope of organic farming

- Increase in awareness and health consciousness
- Global consumers are increasingly looking for organic food, which is considered safe, and hazard free.
- The global prices of organic food are more lucrative and remunerative.
- The potential of organic farming is signified by the fact that the farm sector has abundant organic nutrient resources like livestock, water, crop residue, aquatic weeds, forest litter, urban, rural solid wastes and agro industries, bio-products.
- India offers tremendous scope for organic farming as it has local market potential for organic products

Principles of International Federation of Organic Agriculture Movements- IFOAM, 1972

1. To produce food of high quality in sufficient quantity.
2. To interact in a constructive and life-enhancing way with natural systems and cycles.
3. To consider the wider social and ecological impact of the organic production and processing systems.
4. To encourage and enhance biological cycles within the farming system, involving micro- organisms, soil flora and fauna, plants and animals.
5. To maintain and increase the long-term fertility of soils.
6. To maintain the genetic diversity of the production system and its surroundings, including the protection of wildlife habitats.
7. To promote the healthy use and proper care of water, water resources and all life therein.

8. To use, as far as possible, renewable resources in locally organized production systems.

9. To give all livestock conditions of life with due consideration for the basic aspects of their innate behaviour.

10. To minimize all forms of pollution.

11. To allow every one involved in organic production and processing a quality of life which meets their basic needs and allows an adequate return and satisfaction from their work, including a safe working environment.

12. To progress towards an entire production, processing, and distribution chain which is both socially just and ecologically responsible.

Advantages of organic farming

- Nutrition - Improved soil health makes food dramatically superior in mineral content
- Poison-free - Free of contamination with health harming chemicals like pesticides, fungicides and herbicides.
- Food tastes better
- Food keeps longer - can be stored longer
- Disease and pest resistance - because of healthy plants
- Weed competitiveness - Healthier crops able to compete
- Lower input costs - No costly chemicals used, nutrients are created in-situ (in the farm)
- Drought resistance
- More profitable - Due to greater food value of organic produce consumers are willing to pay premium prices

Disadvantages of organic farming

- Productivity - Low productivity is often reported as the quantum nutrient used comparatively lower
- Labour intensive - Cultivation requires more labour especially for weed control
- Skill - requires considerable skill to farm organically Ex. Choice of alternatives for control of pests

- Lack of convenience in management compared to easier management like fertilizer application in conventional methods

Synonyms of organic farming

- Eco-farming
- Biological farming
- Bio-dynamic farming
- Macrobiotic agriculture

Eco-farming

- Farming in relation to ecosystem.
- It has the potential for introducing mutually reinforcing ecological approaches to food production.
- It aims at the maintenance of soil chemically, biologically and physically the way nature would do it left alone.
- Soil would then take proper care of plants growing on it.
- *Feed the soil, not the plant* is the watchword and slogan of ecological farming.

Biological farming

Farming in relation to biological diversity

Biodynamic farming

Farming which is biologically organic and ecologically sound and sustainable farming.

Components of organic farming: They are i) organic manures, ii) non-chemical weed control measures and iii) biological pest and disease management.

Key principles of organic farming: Organic agriculture systems are based on three strongly interrelated principles. They are 1) Mix farming 2) Crop rotation 3) Organic cycle optimization.

Essential characteristics of organic farming: They are,

i) Maximal but sustainable use of local resources;

ii) Minimal use of purchase inputs, only as complimentary to local resource;

iii) Ensuring the biological functions of soil-water-nutrients-humus continuum;

iv) Maintaining a diversity of plant and animal species as a basis for ecological balance and economic stability.

v) Increasing crop and animal diversity in the form of polycultures, agro forestry systems, Integrated crop / live stock systems etc., to minimize risk.

Options of organic farming

1. *Pure organic farming* is done by the use of organic manures, bio-fertilizers and bio-pesticides and completely avoiding inorganic fertilizers and pesticides.

2. *Integrated green revolution farming* is a high input technology green revolution farming involving INM and IPM. Here chemical fertilizers and pesticides are used apart from organics, bio fertilizers and bio-control agents depending on the necessity.

3. *Integrated Farming System* (IFS) is a resource management strategy to achieve economic and sustained agricultural production through two or more interrelated or inter dependent agricultural and allied enterprises, to meet diverse requirements of the farm household, while preserving the resources base (soil fertility) and maintaining a high environmental quality. It is a low input organic farming (LIOF) in which the local resources are effectively recycled. For example – Cropping (0.96 ha); Fishery (0.04 ha) + poultry in wetlands. Crops, dairy, bio-gas, trees in garden lands. Crops + trees + goats in dry-farming areas.

Integrated Farming System (IFS)

Integration of two or more appropriate combination of enterprises like crop, dairy, piggery, fishery, poultry, bee keeping etc., for each farm according to the availability of resources to sustain and satisfy the necessities of the farmer

Definition: A farming system is a collection of distinct functional units such as crop, livestock, processing, investments and marketing activities which interact because of the joint use of inputs they receive from the environment, which have the common objective of satisfying the farmers' (decision makers) aims. The definition of the borders of the options depends on circumstances; often it includes not only the farm (economic enterprise) but also the household (farm – household system)"

Possible Enterprises

Wetland based farming system	Garden and based farming system	Dry land based farming system
Crop+ Fish+ Poultry / pigeon	Crop + Dairy + Biogas	Crop + Goat + Agroforestry
Crop + Fish + Mushroom	Crop + Dairy + Biogas + Sericulture	Crop + Goat + Agroforestry + Horticulture
	Crop + Dairy + Biogas + Mushroom + Sylvi-culture	

Benefits of IFS

- Higher Productivity
- Profitability
- Sustainability
- Balanced food
- Recycling reduces pollution

- Money round the year
- Employment generation
- Increase input efficiency
- Standard of living of the farmer increased
- Better utilisation of land, labour, time and resources

Dryland Agriculture

- Indian agriculture is predominantly a rainfed agriculture under which both dryfarming and dryland agriculture are included.
- Out of the 143 million ha of total cultivated area in the country, 101 million ha (i.e. nearly 70%) area are rainfed.
- In dryland areas, variation in amount and distribution of rainfall influence the crop production as well as socio-economic conditions of farmers.
- The dryland areas of the country contribute about 42% of the total food grain production.
- Most of the coarse grains like sorghum, pearl millet, finger millet and other millets are grown in drylands only.
- The attention has been paid in the country towards the development of dryland farming.
- Efforts were made to improve crop yields in research projects at Manjari, Solapur, Bijapur, Raichur and Rohtak.
- An all India co-ordinated research project for Dryland Agriculture was launched by ICAR in 1970 in collaboration with Government of Canada and later Central Research Institute for Dryland Agriculture (CRIDA) was established at Hyderabad.

Characteristics of dryland agriculture

Dryland areas may be characterized by the following features,

1. Uncertain, ill-distributed and limited annual rainfall
2. Occurrence of extensive climatic hazards like drought, flood etc
3. Undulating soil surface
4. Occurrence of extensive and large holdings

5. Practice of extensive agriculture, i.e., prevalence of monocropping etc.
6. Relatively large size of fields
7. Similarity in types of crops raised by almost all farmers of a particular region
8. Very low crop yield
9. Poor economy of the farmers

Dryland agriculture

It is the profitable production of useful crops, without irrigation, on lands (arid and semi arid) that receive annual rainfall of less than 750mm

Rainfed agriculture

It is the profitable production of useful crops, without irrigation, on lands (humid & subhumid regions) that receive annual rainfall of more than 750mm

Difference between rainfed and irrigated farming

S.No.	Rainfed farming	Irrigated farming
1	In a certain part of the year crop is grown where rainfall received	Through out the year depending upon the water availability
2	Crops/crop varieties having drought tolerance or less water requirement are used	According to the need, crops or their varieties are selected
3	Duration of crops depends on the rainfall duration/ growing period most of the times short duration (LGP)	Depending upon the need
4	Mixed cropping is beneficial	Generally pure cropping is done
5	Due to limitation of moisture one or two crops in a year is possible	More than two crops in a year are grown, subject to availability of water
6	The field is ploughed to deep to increase infiltration of rains	No need for deep ploughing to conserve soil moisture

Improved dryland technologies

Following are the various improved techniques and practices recommended for achieving the objective of increased and stable crop production in dryland areas.

- **Crop planning**: Crop varieties for dryland areas should be of short duration through resistant tolerant and high yielding which can be harvested within rainfall periods and have sufficient residual moisture in soil profile for post-monsoon cropping.

- **Planning for weather**: Variation in yields and output of the dryland agriculture is due to the observation in weather conditions especially rainfall. An aberrant weather can be categorized in three types viz.,

  a. Delayed onset of monsoon.

  b. Long gaps or breaks in rainfall and

  c. Early cessation of rains towards the end of monsoon season.

  Farmers should make some changes in normal cropping schedule for getting some production in place of total crop failure.

- **Crop substitution**: Traditional crops/varieties which are inefficient utilizers of soil moisture, less responsive to production input and potentially low producers should be substituted by more efficient ones.

- **Cropping systems**: Increasing the cropping intensities by using the practice of intercropping and multiple cropping is the way of more efficient utilization of resources. The cropping intensity would depend on the length of growing season, which in turn depends on rainfall pattern and the soil moisture storage capacity of the soil.

- **Fertilizer use**: The availability of nutrients is limited in drylands due to the limiting soil moisture. Therefore, application of the fertilizers should be done in furrows below the seed. The use of fertilizers is not only helpful in providing nutrients to crop but also, helpful in efficient use of soil moisture. A proper mixture of organic and inorganic fertilizers improves moisture holding capacity of soil and increase during tolerance.

- **Rain water management:** Efficient rain water management can increase agricultural production from dryland areas. Application of compost and farm yard manure and raising legumes add the organic matter to the soil and increase the water holding capacity. The water, which is not retained by the soil, flows out as surface runoff. This excess runoff water can be harvested in storing dugout ponds and recycled to donor areas in the severe stress during rainy season or for raising crops during winter.

- **Watershed management:** Watershed management is an approach to optimize the use of land, water and vegetation in a area and thus, to provide solution drought, moderate floods, prevent soil erosion, improve water availability and increase fuel, fodder and agricultural production on a sustained basis.

- **Alternate Land Use Systems** : All drylands are not suitable for crop production. Same lands may be suitable for range/pasture management and for tree farming and ley farming, dryland horticulture, agro-forestry systems including alley cropping. All these systems which are alternative to crop production are called as alternate land use systems. This system helps to generate off-season employment mono-cropped dryland and also, minimizes risk, utilizes off-season rains, prevents degradation of soils and restores balance in the ecosystem. The different alternate land use systems are alley cropping, agri-horticultural systems and silvi-pastoral systems, which utilizes the resources in better way for increased and stabilized production from drylands.

Questions

Fill in the blanks

1. ____________is a piece of land with specific boundaries, where crop and livestock enterprises are taken up under a common management.
2. The land supplied with water mostly from underground sources is called ______________.
3. Farming in relation to biological diversity is ____________
4. The farming done in areas with an annual rainfall less than 800 mm is __________
5. The production system which avoids or largely excludes the use of synthetic compound fertilizers, pesticides, growth regulators is ______

Choose the correct answer

1. Farming in relation to ecosystem is

   a. Eco-farming b. Biological farming

   c. Bio-dynamic farming d. Organic farming

2. Uncertain, ill-distributed and limited annual rainfall are characteristic features of

   a. Dryands b. Rainfed lands

   c. Wetlands d. Gardenlands

3. In India the area under rainfed farming is

   a. 329 m ha.
   b. 101 m ha.
   c. 42 m ha.
   d. 13 m ha

4. Profitable production of crops without irrigation on lands that receive annual rainfall of more than 750 mm is called as

   a. Dry farming
   b. Dryland farming
   c. Rainfed farming
   d. Irrigated farming

5. Balanced food can be obtained from

   a. Organic farming
   b. Natural farming
   c. Integrated farming system
   d. Dryfarming

Answer the following

1. Dry farming
2. Sustainable Agriculture
3. LEISA
4. Scope of organic farming
5. Integrated farming system.

9

Tillage and Tilth

[illegible]

9

Tillage and Tilth

Tillage operations in various forms have been practiced from the very inception of growing plants. Primitive man used tools to disturb the soils for placing seeds. The word tillage is derived from the Anglo-Saxon words *tilian* and *teolian*, meaning to plough and prepare soil for seed to sow, to cultivate and to raise crops. **Jethro Tull**, who is considered as father of tillage suggested that thorough ploughing is necessary so as to make the soil into fine particles.

Definition

Tillage refers to the mechanical manipulation of the soil with tools and implements so as to create favourable soil conditions for better seed germination and subsequent growth of crops.

Tilth is a physical condition of the soil resulting from tillage. Tilth is a loose friable (mellow), airy, powdery, granular and crumbly condition of the soil with optimum moisture content suitable for working and germination or sprouting of seeds and propagules i.e., tilth is the ideal seed bed.

Characteristics of Good Tilth

Characteristics of good tilth: Good tilth refers to the favourable physical conditions for germination and growth of crops. Tilth indicates two properties of soil viz., the size distribution of aggregates and mellowness or friability of soil. The relative proportion of different sized soil aggregates is known as size distribution of soil aggregates. Higher percentages of larger aggregates with a size above 5 mm in diameter are necessary for irrigated agriculture while higher percentage of smaller aggregates (1-2 mm in diameter) are desirable for dry land agriculture.

Mellowness or friability is that property of soil by which the clods when dry become more crumbly. A soil with good tilth is quite porous and has free drainage upto water table. The capillary and non-capillary pores should be in equal proportion so that sufficient amount of water and free air is retained respectively.

Objectives of Tillage: Tillage is done

1. To prepare ideal seed bed favourable for seed germination, growth and establishment;
2. To loosen the soil for easy root penetration and proliferation;
3. To remove other sprouting materials in the soil;
4. To control weeds;
5. To certain extent to control pest and diseases which harbour in the soil;
6. To improve soil physical conditions;
7. To ensure adequate aeration in the root zone which in turn favour for microbial and biochemical activities;
8. To modify soil temperature;
9. To break hard soil pans and to improve drainage facility;
10. To incorporate crop residues and organic matter left over ;
11. To conserve soil by minimizing the soil erosion;
12. To conserve the soil moisture;
13. To harvest efficiently the effective rain water;
14. To assure through mixing of manures, fertilizers and pesticides in the soil;
15. To facilitate water infiltration and thus increasing the water holding capacity of the soil and
16. To level the field for efficient water management.

Types of Tilth

Fine Tilth refers to the powdery condition of the soil.

Coarse Tilth refers to the rough clody condition of the soil.

Fine seedbed is required for small seeded crops like Ragi, Onion, Berseem, Tobacco.

Coarse seed bed is needed for bold seeded crops like sorghum, cotton, chick pea, lab-lab etc.

Types of Tillages

1. **On Season Tillage**: It is done during the cropping season (June-July or Sept.-Oct.)

2. **Off Season Tillage**: It is done during fallow or non-cropped season (summer)

3. **Special Types of Tillage**: It is done at any time with some special objective/purpose.

1. On Seasonal Tillage

A. Preparatory Tillage

i) Primary tillage

ii) Secondary tillage

iii) Seed bed preparation

B. Inter Tillage

i) Inter cultivation

2. Off Season Tillage

a) Stubble or post harvest tillage

b) Summer tillage (*kodai uzhavu*)

c) Winter tillage

d) Fallow tillage

3. Special Types

i) Sub soil tillage (sub soiling)

ii) Levelling tillage

iii) Wet tillage

iv) Strip tillage

v) Clean tillage

vi) Ridge tillage

vii) Conservation tillage

viii) Contour tillage and

ix) Blind tillage

i) **On Season tillage**: Tillage operations done for raising the crops in the same season or at the onset of the crop season are called as on season tillage. They are,

A. Preparatory tillage

If refers to tillage operations that are done to prepare the field for raising crops. It is divided into three type as,

i) **Primary tillage**: the first cutting and inverting of the soil that is done after the harvest of the crop or untilled fallow, is known as primary tillage. It is normally the deepest operation performed during the period between two crops. Depth may range from 10-30 cm. It includes ploughing to cut and invert the soil for further operation. It consists of deep opening and loosening the soil to bring out the desirable tilth. The main objective is to control weeds to incorporate crop stubbles and to restore soil structure.

ii) **Secondary tillage**: It refers to shallow tillage operation that is done after primary tillage to bring a good soil tilth. In this operation the soil is stirred and conditioned by breaking the clods and crust, closing of cracks and crevices that form on drying. Incorporation of manures and fertilizers, levelling, mulching, forming ridges and furrows are the main objectives. It includes cultivating, harrowing, pulverizing, rakking, levelling and ridging operations.

iii) **Seed bed preparation**: Refers to a very shallow operation intended to prepare a seed bed or make the soil to suit for planting. Weed control and structural development of the soil are the objectives.

B. Inter tillage

i) Refers to shallow tillage operation done in the field after sowing or planting or prior to harvest of crop plants i.e. tillage during the crop stand in the field. It includes inter cultivating, harrowing, hoeing, weeding, earthing up, forming ridges and furrows etc., Inter tillage helps to incorporate top dressed manures and fertilizers, to earth up and to prune roots.

ii) **Off Season tillage:** Tillage operation is done for conditioning the soil during uncropped season with the main objective of water conservation, leveling to the desirable grade, leaching to remove salts for soil reclamation reducing the population of pest and diseases in the soils. etc., They are,

a) **Stubble or Post harvest tillage**: Tillage operation carried out immediately after harvest of crop to clear off the weeds and crop residues and to restore the soil structure. Removing of stiff stubbles of sugarcane crop by turning and incorporating the trashes and weeds thus making the soil ready to store rain water etc., are the major objectives of such tillage operations.

b) **Summer tillage (Kodai Uzhavu)**: Operation being done during summer season in tropics to destroy weeds and soil borne pest and diseases, checking the soil erosion and retaining the rain water through summer showers. It affects the soil aggregates, soil organic matter and sometimes favour wind erosion.

c) **Winter tillage**: It is practiced in temperate regions where the winter is severe that makes the field unfit for raising crops. Ploughing or harrowing is done in places where soil condition is optimum to destroy weeds and to improve the physical condition of the soil and also to incorporate plant residues.

d) **Fallow tillage**: Refers to the leaving of arable land uncropped for a season or seasons for various reasons. Tilled fallow represent an extreme condition of soil disturbance to eliminate all weeds and control soil borne pest etc., Fallow tilled soil is prone to erosion by wind and water and subsequently they become degraded and depleted.

iii) Special types of tillage

a) **Subsoil tillage (subsoiling)** is done to cut open/break the subsoil hard pan or plough pan using sub soil plough/chisel plough. Here the soil is not inverted. Sub soiling is done once in 4-5 years, where heavy machinery is used for field operations and where there is a colossal loss of top soil due to carelessness. To avoid closing of sub soil furrow vertical mulching is adopted.

b) **Levelling by tillage:** Arable fields require a uniform distribution of water and plant nutrition for uniform crop growth. This is achieved when fields are kept fairly levelled. Levellers and scrapers are used for levelling operations. In levelled field soil erosion is restricted and other management practices become easy and uniform.

c) **Wet tillage:** This refers to tillage done when the soil is in a saturated (anaerobic) condition. For example puddling for rice cultivation.

d) **Strip tillage:** Ploughing is done as a narrow strip by mixing and tilling the soil leaving the remaining soil surface undisturbed.

e) **Clean tillage:** Refers to the working of the soil of the entire field in such a way no living plant is left undisturbed. It is practiced to control weeds, soil borne pathogen and pests.

f) **Ridge tillage:** It refers to forming ridges by ridge former or ridge plough for the purpose of planting.

g) **Conservation tillage:** It means any tillage system that reduces loss of soil or water relative to conventional tillage. It is often a form of non-inversion tillage that retains protective amounts of crop residue mulch on the surface. The important criteria of a conservation tillage system are, i) presence of crop residue mulch, ii) effective conservation of soil and water, iii) improvement of soil structure and organic matter content and iv) maintenance of high and economic level of production.

h) **Contour tillage:** It refers to tilling of the land along contours (contour means lines of uniform elevation) in order to reduce soil erosion and run off.

i) **Blind tillage:** It refers to tillage done after seeding or planting the crop (in a sterile soils) either at the pre-emergence stage of the crop plants or while they are in the early stages of growth so that crop plants (cereals, tuber crops etc.) do not get damaged, but extra plants and broad leaved weeds are uprooted.

Factors Affecting (Intensity and Depth of) the Tillage Operations

Several factors are responsible for deciding depth and intensity and depth of tillage operations. They are soil type, crop and variety, type of farming moisture status so the soil, climate and season, extent of weed infestation, irrigation methods, special needs and economic condition, and knowledge and experience of the farmer. Prominent factors are,

i) *Crop*: It decides the type, intensity and depth of tillage operations with small sized seeds like fingermillet, tobacco etc., Require a fine seedbed which can provide intimate soil-seed contact as against coarser seed bed required for larger size seeds such as sorghum, maize, pulses, etc., Root or tuber crops require deep tillage whereas rice requires shallow puddling.

ii) *Soil type*: It dictates the time of ploughing. Light soils require early and rapid land preparation due to free drainage and low retentive capacity as against heavy soils.

iii) *Climate*: It influences soil moisture content; draught required tilling and the type of cultivation. Low rainfall and poor water retentive capacity of shallow soil do not permit deep ploughing at the start of the season. Heavy soils developing cracks during summer (self tilled) need only harrowing. Light soils of arid regions need coarse tilth to minimize wind erosion.

iv) *Type of farming*: It influences the intensity of land preparation. In dry lands, deep ploughing is necessary to eradicate perennial weeds and to conserve soil moisture. Repeated shallow tilling is adequate under such intensive cropping.

v) *Cropping system*: In involves different crops which need different types of tillage. Crop following rice needs repeated preparatory tillage for obtaining an ideal seed-bed. Crops following tuber crops like potato require minimum tillage. Similarly crops following pulses need lesser tillage than that of following sorghum, maize or sugarcane.

Depth of Ploughing

Desirable ploughing depth is 12.5 to 20 cm. Ploughing depth varies with effective root zone depth of the crops. Ploughing depth is 10-20 cm to shallow rooted crops and 15-30 cm to deep rooted crops. Deep ploughing is done to control perennial weeds like *Cyanodon dactylon* and to break soil hard pans. Since deep ploughing increases the cost, most farers resort to shallow ploughing only.

Number of ploughing: It depends on soil conditions, time available for cultivation between two crops, (turn over period) type of cropping systems etc. Small or fine seeded crop requires fine tilth, which may require more ploughing. Zero tillage is practiced in rice fallow pulse crops or relay cropping system. Three numbers of puddling is sufficient for rice cultivation. Minimum number of ploughing are taken up at optimum moisture level to bring favourable tilth depending on the need of the crop and financial resources of the farmer. In fact, this brought the concept of minimal tillage or zero tillage systems.

Time of ploughing (Based on moisture status and soil type): The optimum moisture content for tillage is 60% of field capacity. Ploughing at right moisture content is very important. Summer ploughing (March – May) can be practiced utilizing summer showers to control weeds and conserve soil moisture. Light soils can be worked under wide range of moisture. Loamy soils can be easily brought to good tilth. Pulverization of clay soils is difficult as they dry into hard clods.

Method of ploughing: Ploughing aims at stirring and disturbing the top layer of soil uniformly without leaving any unploughed strips of land. Straight and uniformly wide furrows give a neat appearance to the ploughed field. When the furrows are not straight or when the adjacent furrows are not uniformly spaced, narrow strips of land are left unploughed. The correct inter furrow space is little over the width of the furrow slice.

After the harvest of a crop the land is first ploughed along the length of the field. This reduces the number of turns at the head lands for opening fresh furrows. The next ploughing is done across the field for breaking furrows of the previous ploughing. This must increase the turns at the headlands and the empty turns along the head lands, but is unavoidable. New turns are taken 6 m wide each time, till the entire field is covered.

Modern Concepts in Tillage

- Conventional tillage involves primary tillage to break open and turn the soil followed by secondary tillage to obtain seed bed for sowing or planting.
- With the introduction of herbicides in intensive farming systems, the concept of tillage has been changed.
- Continuous use of heavy ploughs create hard pan in the subsoil, results in poor infiltration.
- It is more susceptible to run-off and erosion.
- It is capital intensive and increase soil degradation.
- To avoid these ill effects, modern concepts on tillage is followed.

1. Minimum tillage

It aims at reducing tillage operations to the minimum necessity for ensuring a good seed bed.

The advantages of minimum tillage over conventional tillage are,

- The cost and time for field preparation is reduced by reducing the number of field operations.
- Soil compaction is comparatively less.
- Soil structure is not destroyed.
- Water loss through runoff and erosion is minimum.
- Water storage in the plough layer is increased.

Tillage can be reduced in 2 ways :

1. By omitting operations which do not give much benefit when compared to the cost.
2. By combining agricultural operations like seeding and fertilizer application.

The minimum tillage systems can be grouped into the following categories,

a) Row zone tillage
b) Plough plant tillage
c) Wheel track tillage

a. Row zone tillage

Primary tillage is done with mould board plough in the entire area of the field; secondary tillage operations like discing and harrowing are reduced and done only in row zone.

b. Plough plant tillage

After the primary tillage, a special planter is used for sowing. In one run over the field, the row zone is pulverized and seeds are sown by the planter

c. Wheel track tillage

Primary ploughing is done as usual. Tractor is used for sowing; the wheels of the tractor pulverize the row zone in which planting is done.

In all these systems, primary tillage is as usual. However, secondary tillage is replaced by direct sowing in which sown seed is covered in the row zone with the equipment used for sowing.

2. Zero tillage (No tillage)

In this, new crop is planted in the residues of the previous crop without any prior soil tillage or seed bed preparation.

It is possible when all the weeds are controlled by the use of herbicides.

Zero tillage is applicable for soils with

- A coarse textured surface horizon,

- Good internal drainage, high biological activity of soil fauna,
- Favourable initial soil structure and
- An adequate quantity of crop residue as mulch.

These conditions are generally found in *Alfisols, Oxisols* and *Ultisols* in the humid and sub-humid tropics.

Till planting

- Till planting is one method of practicing zero tillage.
- A wide sweep and trash bar clears a strip over the previous crop row and planter opens a narrow strip into which seeds are planted and covered.
- Here, herbicide functions are extended.
- Before sowing, the vegetation present has to be destroyed for which broad spectrum non selective herbicides like glyposate, paraquat and diquat are used.

Advantages

- Zero tilled soils are homogenous in structure with more number of earthworms.
- Organic matter content increases due to less mineralization.
- Surface run-off is reduced due to presence of mulch.

Disadvantages

- Higher amount of nitrogen has to be applied for mineralization of organic matter in zero tillage.
- Perennial weeds may be a problem.
- High number of volunteer plants and buildup of pests.

3. Stubble mulch tillage or stubble mulch farming

- Soil is protected at all times either by growing a crop or by leaving the crop residues on the surface during fallow periods.
- Sweeps or blades are generally used to cut the soil up to 12 to 15 cm depth in the first operation after harvest and depth of cut is reduced during subsequent operations.

- When large amount of residues are present, a disc type implement is used for the first operation to incorporate some of the residues into the soil.
- This hastens the decomposition but still keeps enough residues on top soil.

Disadvantages of stubble mulch farming

- The residues left on the surface interfere with seed bed preparation and sowing operations.
- The traditional tillage and sowing implements or equipments are not suitable under these conditions.

4. Conservation tillage

- The major objective is to conserve soil and soil moisture.
- It is a system of tillage in which organic residues are not inverted into the soil such that they remain on surface as protective cover against erosion and evaporation losses of soil moisture.
- If stubble forms the protective cover on the surface, it is usually referred to as stubble mulch tillage.
- The residues left on soil surface interfere with seed bed preparation and sowing operations.
- It is a year round system of managing plant residue with implements that undercut residues, losses the soil and kills the weeds.

Advantages

- Energy conservation through reduced tillage operations.
- Improve the soil physical properties.
- Reduce the water runoff from fields.

Main field preparation

- Tillage operations are generally classified in to two, preparatory cultivation and after cultivation.
- The preparatory cultivation or tillage is operations that are done before the cultivation.
- This preparatory cultivation is generally called as main field preparation.

- The main field preparation involves three processes, viz., primary tillage, secondary tillage and lay-out for sowing.
- Some of the important primary tillage implements are country plough, mould board plough, disc plough, chisel plough etc.
- Cultivators and harrows are generally used for secondary tillage purpose.
- However, in practical means, the first two (primary and secondary tillages) may not have any key difference, since; both operations are mainly carried out with same implement.
- Country plough and cultivators are used for both the purposes.
- After thorough ploughing, the field modified in to suitable way for planting such as ridges and furrows or beds and channels or pit according to the need of the crops.
- Such field modifications are called land configurations.

Tillage Implements

Any device used to carry on some work is called as implement. Implements are operated by animal power or by machinery. Implements are classified into primary, secondary and intercultural depending on the purpose for which it is being used.

Primary Tillage Implements

Primary tillage is the deepest operations performed between two crops.The following are the implements used for primary tillage.

Primary Tillage Implements

1. Country plough
2. Improved iron plough
3. Bose plough
4. Mould board plough
5. Turnwrest plough
6. Disc plough
7. Reversible disc plough
8. Chisel plough

1. Country / wooden / Desi plough

The indigenous plough consists of a wooden body to which a handle and a shaft pole are attached. The body is made of a bent piece of hard wood with two arms making an angle of about 135^0. It is given a wedge shape with an isosceles triangular section. A small piece of flat iron (shares) serves as the piercing point of the plough and is fixed over the plough body with clamps. The shaft pole is secured with the yoke during working. The working of plough results in the opening of 'V' shaped furrow. The width of furrow depends on the size of the plough bottom. (fig. 1)

The depth of penetration of a country plough can be altered a little by pulling the implement behind or pushing forward which results in deeper or shallow ploughings respectively. It covers 0.15 to 0.20 hectare in 8 hours.

2. Improved iron plough

The bullock drawn improved iron plough is made of mild steel except the poleshaft and hence it has longer life. As and when the share wears off, it can be pushed forward. Poleshaft angle and height of the handle can be adjusted according to field requirements. The plough is provided with a mould board as optional attachment for soil inversion. This plough is suitable for dry ploughing in all types of soil with a pair of bullocks. It covers 0.5 ha per day and costs Rs. 500/-. (fig. 2)

3. Bose plough (Melur plough)

It is wooden plough with a mould board can share instead of the usual small iron share. It is used for the primary tillage operations in wetlands. Nowadays this plough is made up of iron angles instead of woods to make it sturdy. The cost of the plough is Rs. 200/-. (Fig. 3)

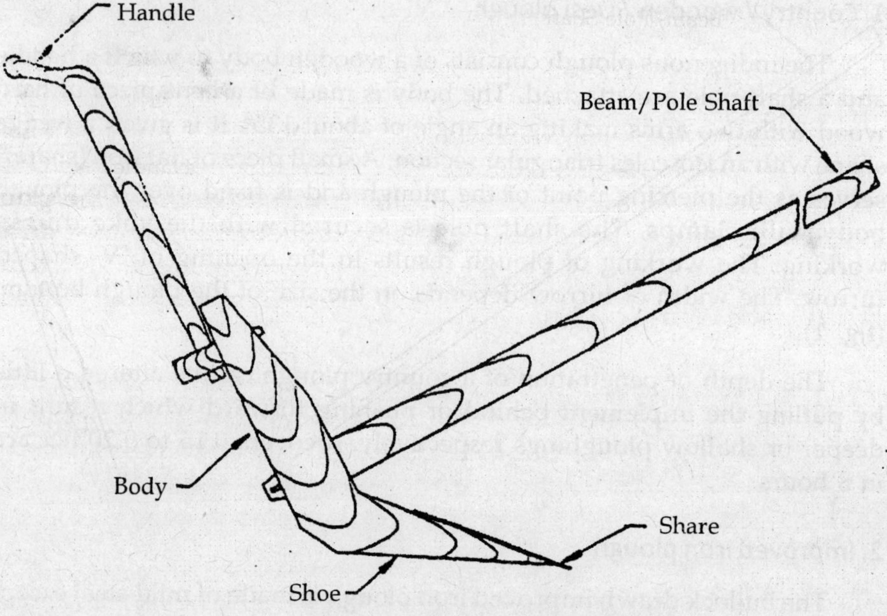

Fig. 1. Country Plough

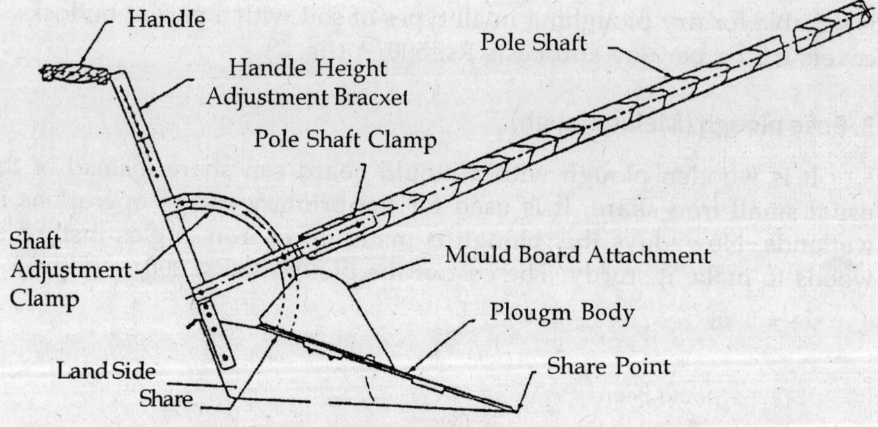

Fig 2. Improved Iron Plough

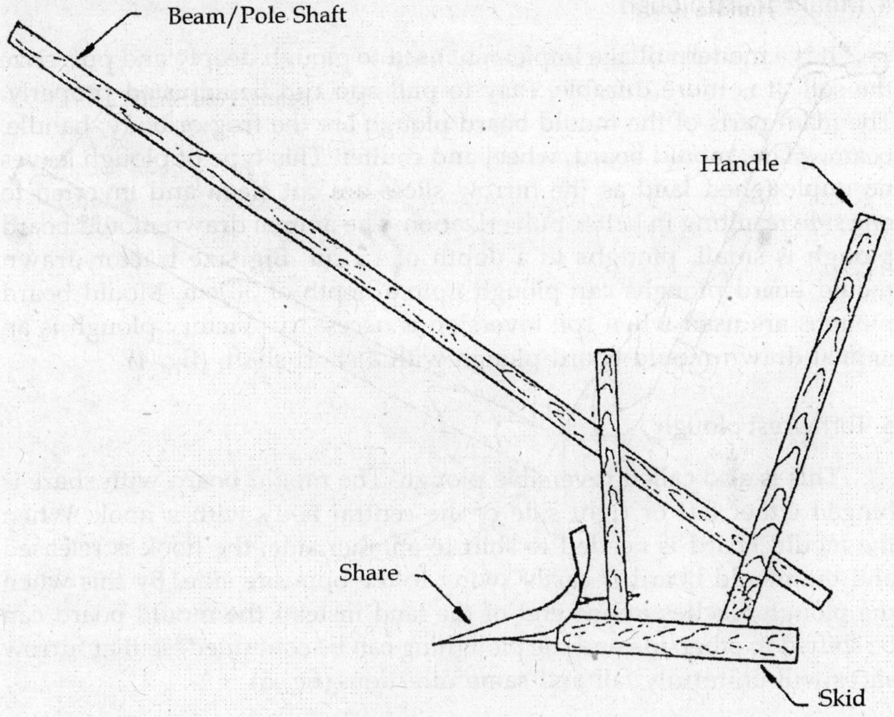

Fig. 3. Melur Plough

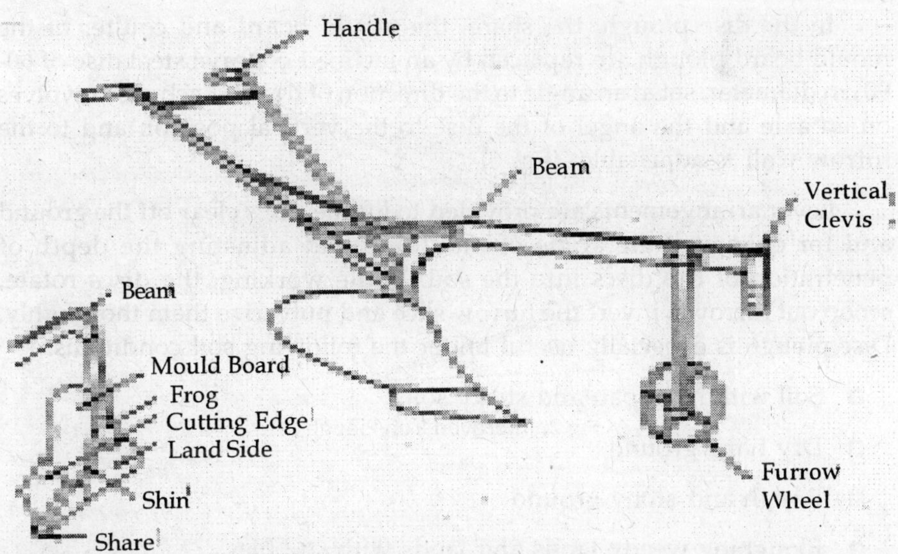

Fig 4. Mould Board Plough

4. Mould board plough

It is a modern tillage implement used to plough deeply and pulverize the soil. It is more durable, easy to pull and can be adjusted properly. The main parts of the mould board plough are the frog or body, handle, beam, share, mould board, wheel and coulter. This type of plough leaves no unploughed land as the furrow slices are cut clean and inverted to one side resulting in better pulverization. The animal drawn mould board plough is small, ploughs to a depth of 15 cm. Big size tractor drawn mould board ploughs can plough upto a depth of 30 cm. Mould board ploughs are used when soil inversion is necessary. Victory plough is an animal drawn mould board plough with a short shaft. (fig. 4)

5. Turnwrest plough

This is also called reversible plough. The mould board with share is hinged either left or right side of the central body with a hook. When the mould board is needed to shift to another side, the hook is released and the mould board is easily swing to the opposite side. By this when the plough reaches to one end of the land instead the mould board can be shifted to otherside and the ploughing can be continued, so that furrow slices will uniformly fall and same direction. (fig. 5)

6. Disc plough

In the disc plough, the share, the mould board and coulter of the mould board plough are replaced by an inclined concave steel disc of 60-90 cm diameter, set at an angle to the direction of travel. Each disc revolves on an axle and the angel of the disc to the vertical position and to the furrow wall is adjustable. (fig. 6)

Lever arrangements are provided to lift the discs clear off the ground and for changing the angle of moulding and adjusting the depth of penetration of the discs into the soil. While working, the discs rotate, scoop out furrows, invert the furrow slice and pulverize them thoroughly. Disc plough is especially useful under the following soil conditions.

- Soil with hard pan and sticky soil,
- Dry hard ground
- Rough and stony ground
- Ploughing weedy lands and lands with stubbles
- Deep ploughing

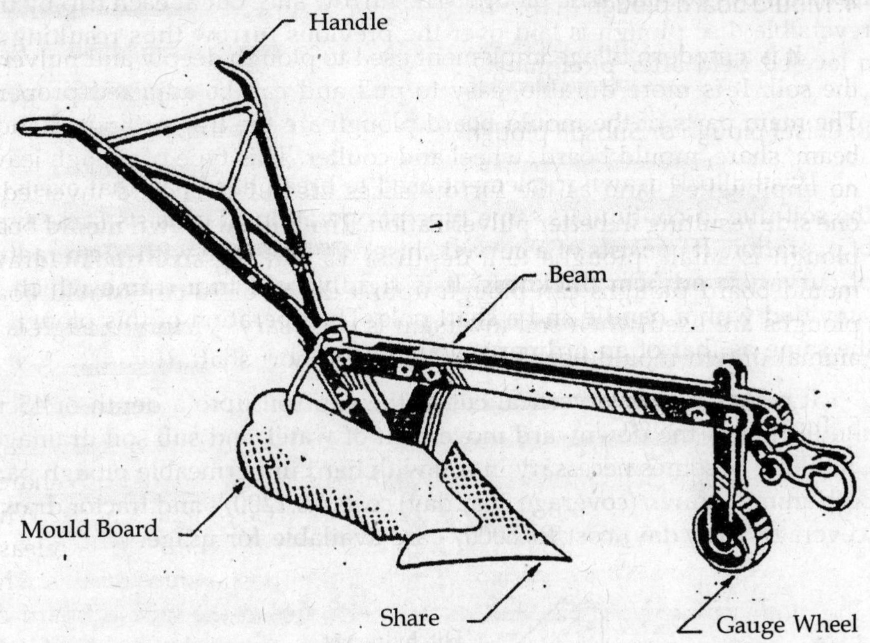

Fig 5. Turn Wrest Plough

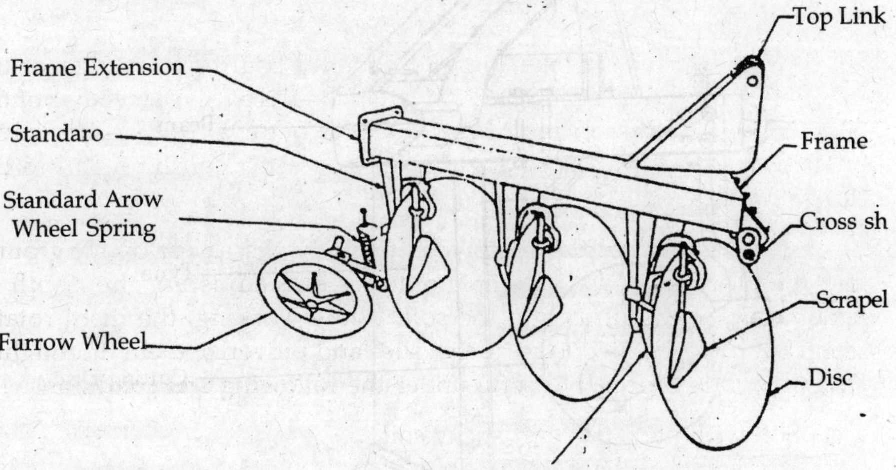

Fig. 6. Disc Plough

7. Reversible disc plough

It is constructed in such a way that the disc can be reversed and the soil is thrown on one side. The land and furrow wheel adjust themselves properly when the plough is reversed. Reversible disc plough saves time taken up by ordinary disc plough. The furrow slice cut at each trip by the reversible disc plough is laid over the previous furrow thus resulting in a leveled field after ploughing.

8. Chisel plough or Subsoil plough

It is bullock drawn implement used to break hard pan that exists in the soil due to continuous same type of operation. It consists same type of operation. It consists of a curved chisel 'C" like tyne with 37cm radius of curvature ad 3cm thickness. It is rigidly held in a frame which is provided with a handle and a shaft pole. The operation of this plough is the same as that of an ordinary plough. (fig. 7)

It makes a simple vertical cut in the sub soil upto a depth of 45cm and fecilitates the downward movement of water and sub soil drainage. Chiselling becomes necessary in soil with hard impermeable plough pan. Both animal drawn (coverage 2ha/ day) costs Rs.1200/- and tractor drawn (coverage 5 ha/day) cost Rs.6000/- are available for usage.

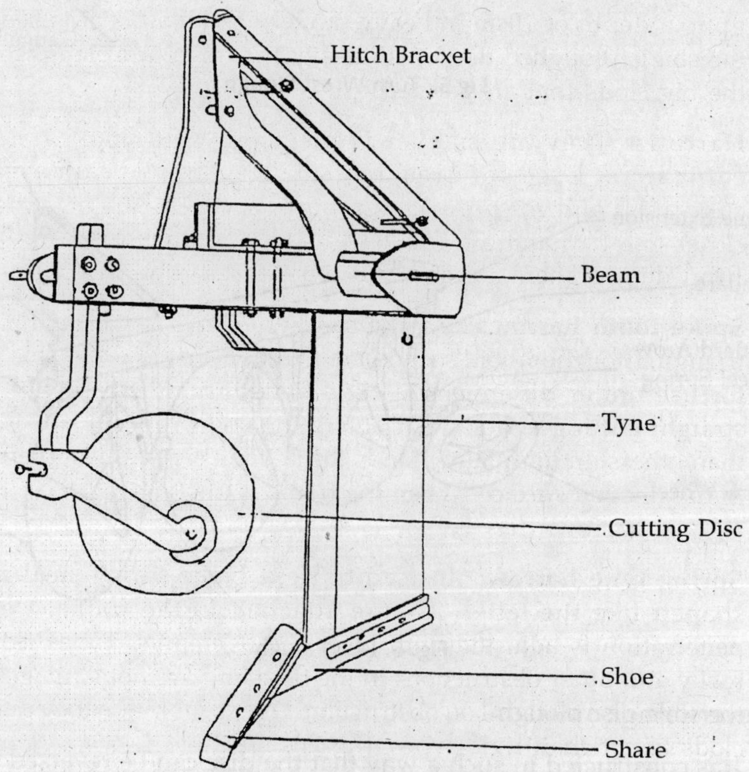

Fig. 7. Subsoil Plough

Secondary Tillage Implements

Secondary tillage is the shallow operation performed after the primary tillage. Secondary tillage implements are used for breaking clods and producing a loose, friable, smooth state. These implements are used with the following objectives.

- Breaking the furrow slice and working the soil to get the required tilth
- Destruction of weeds
- Strirring the soil and forming mulch
- Mixing the manures and fertilizers with soil
- Covering the seeds

1. **Cultivators:** These implements have number of tynes for piercing the soil and breaking clods. Tynes of 23-30 cm long are fixed to a heavy and sturdy, frame, mounted on wheels. These tines penetrate up to a depth of 20cm in heavy models. Cultivators are used when the soil is ploughed deep with heavy mould board plougs to break the big clods that are formed.

2. **Harrows:** They are smaller implements with many tynes like cultivators. Used for breaking smaller clods left unbroken by cultivators and for producing a powdery seed bed. Tynes are set closer (5-8 cm) and are smaller in size. They penetrate upto about 10cm depth. There are different types of harrows in use.

A) **Spike tooth harrow:** Peg like steel tynes of round, oval, square, triangular or rhomboid section are fixed on a rigid or flexible frames for use under different soil conditions. Rhomboid section offers straight cutting edge and it enters the soil properly and is better than others. In undulating lands the flexible types adjust themselves to the uneven surface. When the frame is of a zig-zag type it called zig-zag harrow. (fig. 8)

B) **Spring tyne harrow:** Instead of rigid tynes strong steel springs shaped like the letter "C" are attached to the frame. Depth of penetration is adjusted with lever arrangements. Tynes ride over rocky and other obstructions in the field and are not damaged since the spring tynes recoil on obstruction. Due to vibration they pulverize clods better than rigid types. (fig. 9)

C) **Chain harrow:** Number of stout steel links are connected together to spread over the soil like a mat. Links may have spike like projections. Since, they are flexible they adjust unevenness of the surface. These harrows are used for breaking clods and making the surface smooth and even. It can also be used for covering seeds after broadcasting.

D) **Disc harrow:** These harrows are made up of number of concave discs of 46-56 cm in diameter, fitted 15cm apart on square axles. Two sets of discs are mounted on different axles. Discs cut through the soil and effectively pulverize clods. Small animal drawn harrows have six discs and power driven harrows have larger number. (fig. 10)

E) **Intercultivating harrow**: Different types of harrows are used for intercultivation. Tynes pass through the inter row spaces and effectively remove the weeds. The typical example of the intercultivating harrow is junior hoe as shown in fig. 7.

Different attachments can be made in the tynes of the junior hoe to make use of same harrow for different purposes.

1. *Sweeps*: are blades that move horizontally under the soil and cut the shallow rooted weeds. There are two kinds of sweeps.

   The *Central sweep* attached to the central tynes has horizontal wings extending on both the sides. The *One side sweep* has the wing on the right or leftside. On the side tynes, one sided sweeps are fixed on the side away from the crop rows.

2. *Hiller* is a rhomboidal curved steel plate, shaped like the mould board which is used for earthing up crop rows.

3. *Furrows* have a double mould board one on either side which splits the furrow slice and lays it on both sides equally. It is used with a central tine to open the furrow for sowing or to clean the furrow for irrigation.

4. *Cultivator* steel is a steel plate with sharp edge which penetrates into the soil.

F) **Blade harrows**: Different from conventional harrows in that there are no tines but they are fixed with horizontal blades which enter into the soil and travel below the surface at a constant depth. These blades severe (cut) the surface layer from the soil below and leaves it in its original position with slight disturbance to the surface soil. These harrows cut the weeds, eradicate all weeds except those which have under ground bulbs. (fig. 11)

The Guntaka is the blade harrow used for primary tillage in ceded districts of Andhra Pradesh. It has a horizontal wooden beam of 15 cm diameter with a fixed handle, shaft pole and blade. The blade is fixed to the beam near the ends by two standards at 25 cm distance from beam. The blade is 1.0m long, 7.5 cm broad and 1.25 cm thick, with a cutting edge in front. Big sized guntakas are called as *bara guntakas* (1.8m long blade). Small guntakas with 15 to 33 cm long blades are called as *danties* which are used for inter cultivation in crops, spaced at 28-46 cm apart. Since they are small, five or six danties are attached to a common yoke and guided by three or four people. It covers 0.4ha/ day of eight hours. (fig. 12)

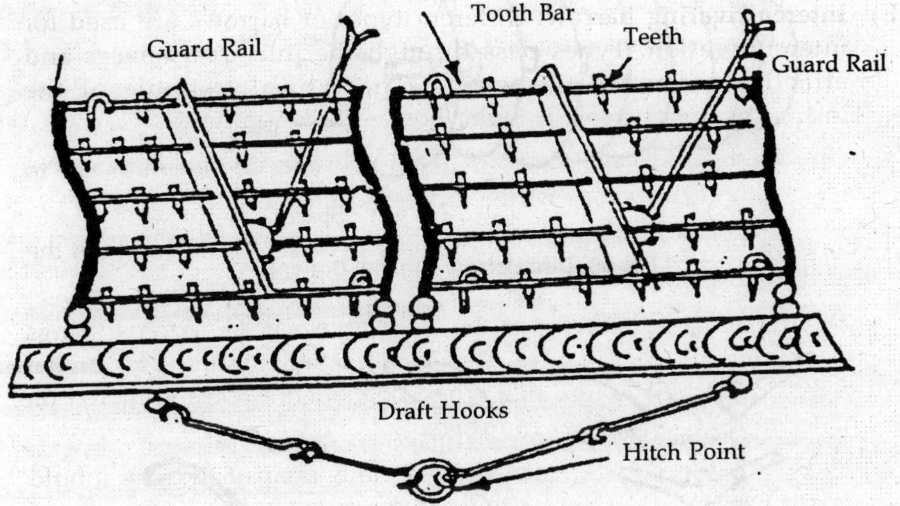

Fig. 8. Spike Tooth Harrow

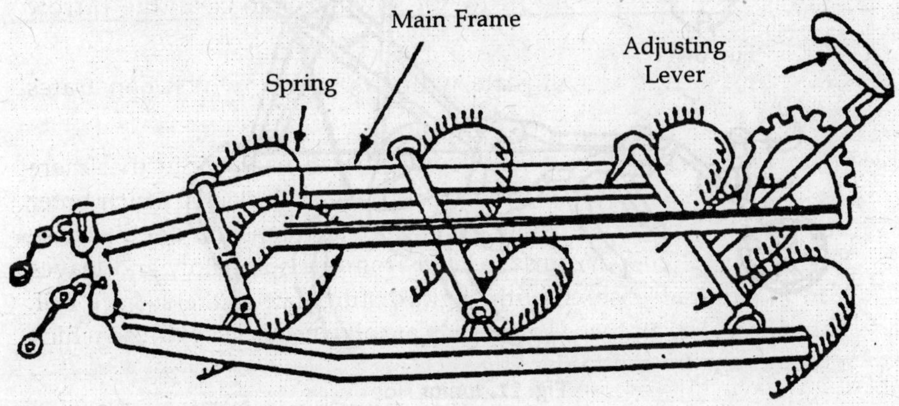

Fig. 9. Spring Tooth Harrow

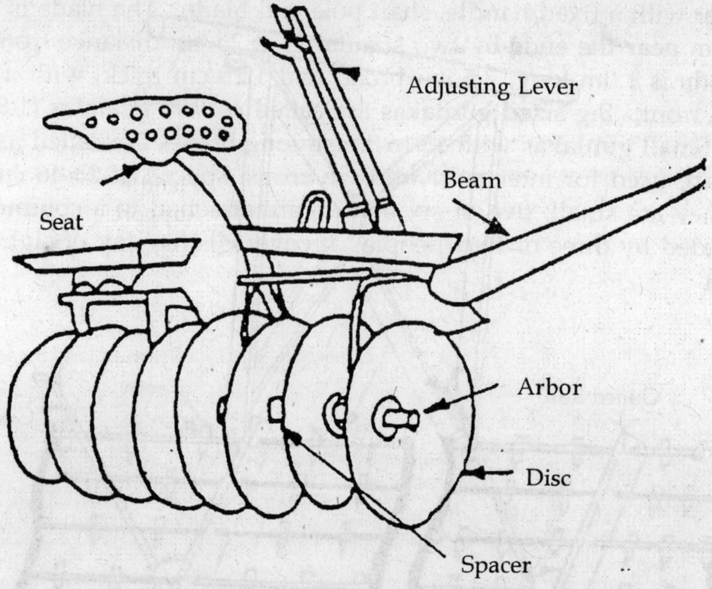

Fig. 10. Disc Harrow (Animal Drawn)

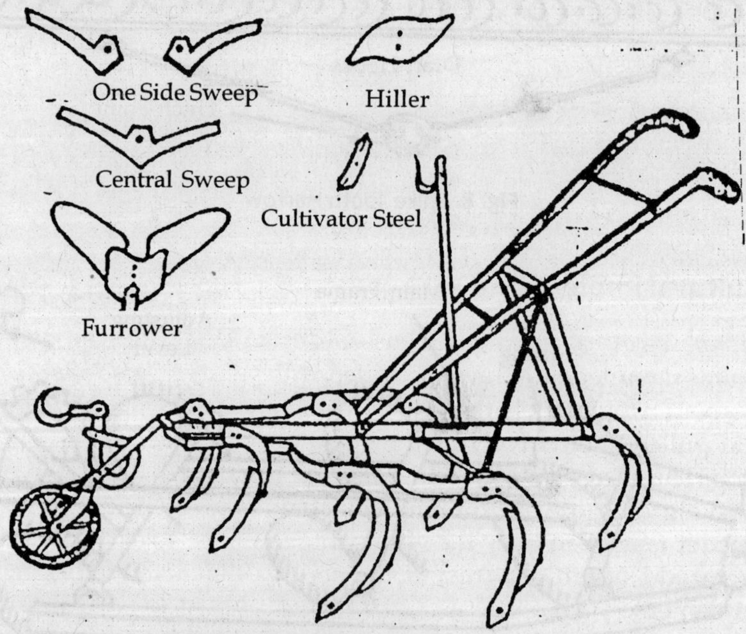

Fig. 11. Junior Hoe

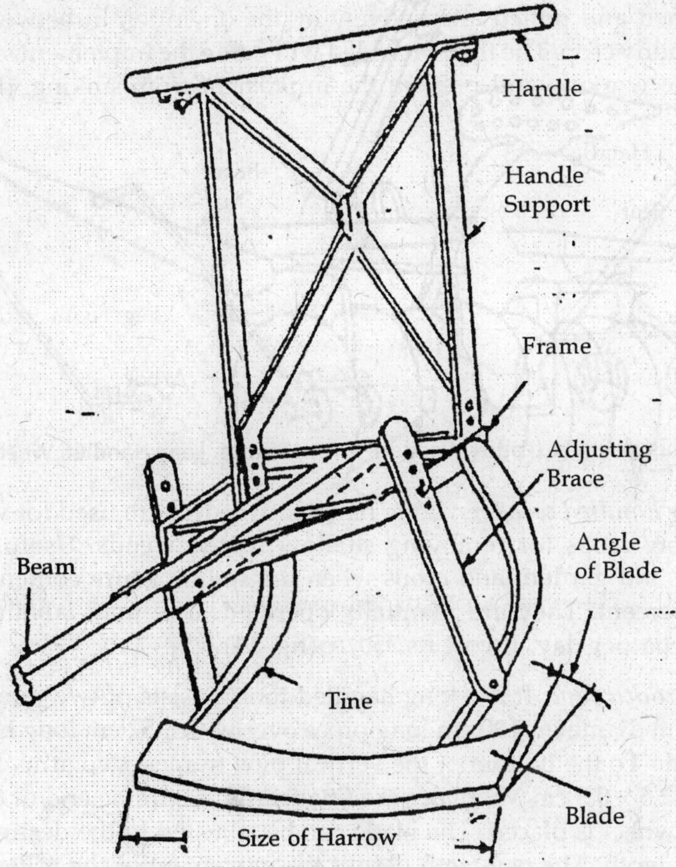

Fig. 12. Blade Harrow (Guntaka)

Inter Cultural Implements

i) *Japanese rotary weeder:* It consists of two small toothed rollers or drums mounted on a frame provided with handle. Each roller consists of about 5 toothed blades. This implement, while working is pushed and pulled alternatively by the operator in between rows of paddy crop. The float provided will guide the implements smoothly while working and prevent the implement sinking into the puddle. The weeder is used to bury the weeds into the mud so as to decompose them add organic matter to the soil, sufficient for working this implement.

ii) *Cono weeder*: It is also similar to rotary weeder in which instead of two toothed rollers or drums two toothed cones are mounted an a frame provided with handle. This implement while working is pushed and pulled alternatively by the operation in between rows of paddy crop. The float provided will guide the implement smoothly while working and prevent the implement from sinking. (fig. 13)

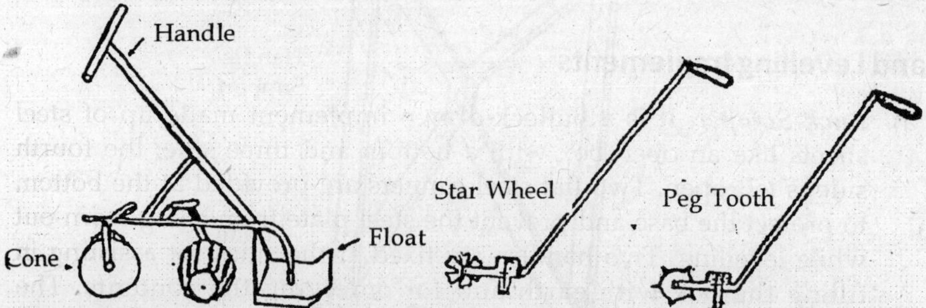

Fig. 13. Cono Weeder (Paddy Weeder) **Fig. 14.** Long Handled Weeders

iii) ***Long handled weeders***: Long handled weeders are used for weeding in row crops for removing shallow rooted weeds. Useful in dry land and garden land crops when the soil moisture content is 8 to 10 percent. They are manually operated. One man labour covers 0.05 ha per day. It cost Rs.350/-. (fig. 14)

a) ***Peg tooth type***: It is a long handled tool consists of two numbers of 2.5 cm diameter, 120 cm long pipes over which 52 cm long handle is fitted. To the bottom of the vertical pipe frames, two arms made of 25 x 2.5 x 0.3 cm MS plates are fitted. At the extreme end of the arm, peg wheel is placed. The blade can be adjusted to the desired angle and depth. The peg teeth permit the movement of the roller in clay soil without getting clogged.

b) ***Star wheel type***: It is similar to the peg type weeder excepting that the star type roller facilitates easier operation of the weeders in loamy and sandy soils.

Special Purpose Implements

Implements that are used for a specific purpose other than primary or secondary and intercultural tillage are called as special purpose implements. The following are some of them.

Multipurpose tool bar / carrier

The multipurpose tool carrier is used for primary, secondary and intercultural operations, forming bunds, ridges and furrows and for sowing crops in rows. It is suitable for all soils. A multipurpose tool carrier is made up to G.I. tube which has the provision to attach cultivators

(4 nos.), ploughs (3 Nos.), ridger (2 Nos.), seed drill (4 Nos.) and bund formers (2 Nos.). The spacing between rows are adjustable. The field capacity of plough is 1.2 ha/day, while for cultivator and ridger it is about 0.74 ha/day.

Land Levelling Implements

a) ***Buck Scraper***: It is a bullock-drawn implement made up of steel sheets like an open box with a bottom and three side; the fourth side is left open. Two flat steel runners are provided at the bottom to protect the base and prevent the steel plate from being worn-out while levelling. Two handles are fixed at the sides for assisting in filling the box with earth and for emptying the contents. The drawbar is attached to the sides with hinge arrangement. It is a bullock drawn implement very useful to carry the soil to a long distance while levelling.

b) ***Levelling board***: It is a channel like or trapezoidal shaped wooden board with 2 to 2.75 m length and 20 cm diameter. It is attached to the shaft pole with a hinged hook. It is used for levelling rice fields after the final ploughing to facilitate uniform seed germination. When the operator stands over the board; it sinks lightly into the loose mud and when it moves, the soil infront of it is also moved, but is released when he gets down. (Fig. 15)

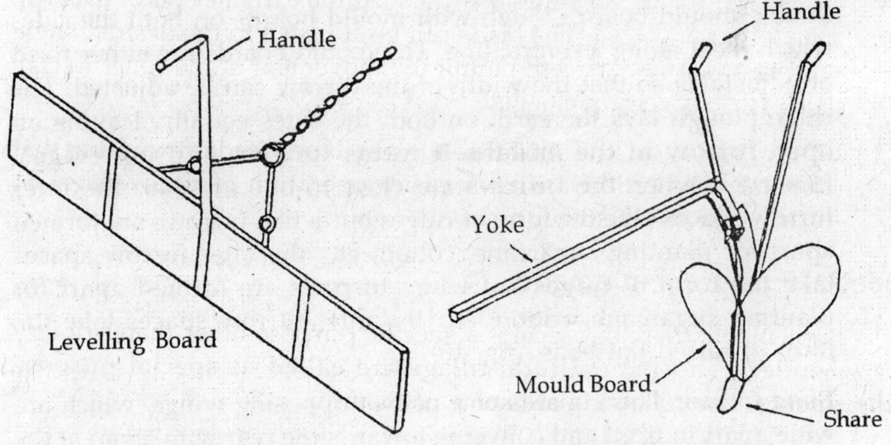

Fig 15. Leveller

Fig 16. Ridge Plough

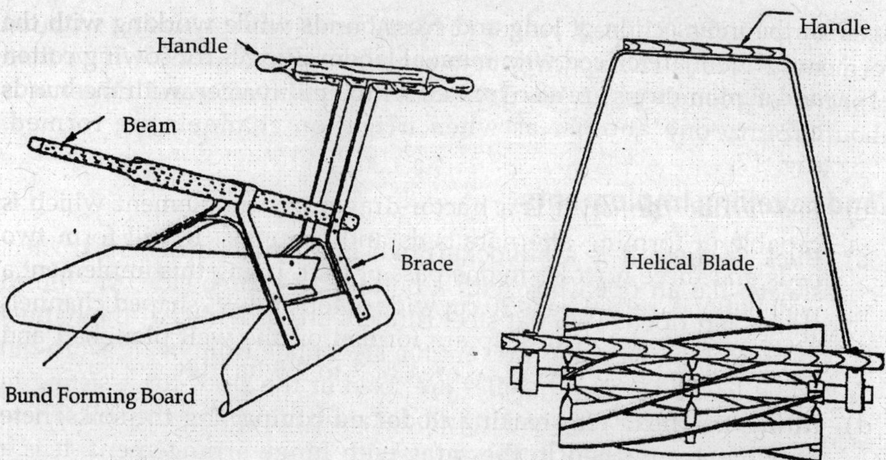

Fig 17. Bund Former **Fig 18.** Helical Blade Puddler

c) ***Wooden float***: It is a bullock drawn implement with a long sledge-like drag used for land smothering. By working the field with wooden float three or four times lengthwise, crosswise and diagonally the field is smoothen in a better way.

Land Shaping Implements

a) ***Ridge plough***: It is a bullock or tractor drawn implement. It is a double mould board plough with mould boards on both the sides which meet along a central line. The mould boards are either fixed or adjustable so that the width of the furrow can be adjusted. The ridge plough lays the earth on both the sides equally, leaving an open furrow in the middle. It forms furrows and not ridges. However, when the furrows are close to one another; the inter furrow spaces take the form of ridges but when furrows are formed apart for planting sugarcane, cotton, etc. the inter furrow spaces take the form of ridges but when furrows are formed apart for planting sugarcane, cotton, etc. the inter-furrow spaces take the form of raised flat beds. (fig. 16)

b) ***Bund former***: This consists of a pair of opposing wings, which are wide apart in front and converge towards the rear with a gap at the end. The wings gather loose soil from the surface and leave it in the form of bund with approximately 18 to 20 cm in height. To form reduced size bunds, the wings are lightly raised at the fore end and pulled a little backward. When the implement is hitched near the yoke, a small quantity of earth alone is gathered. The gaps formed at the intersection of long and cross bunds while working with the implement are closed with manual labour. Ridges for sowing cotton and similar crops are also formed with bund former, with the bunds close to one another as when irrigation channels are formed. (fig. 17)

c) *Bed-furrow former*: It is a tractor-drawn iron implement which is capable of forming alternate beds and channels. It will form two beds and three furrows in one pass of unit. Using this implement a well defined, raised beds 30 cm wide at top and 'V' shaped channels 45cm wide and 15 cm deep are formed on the well ploughed and harrowed field. It covers an area of 3 to 3.5 ha/day.

d) ***Rollers*** are used for breaking clods and compacting the soil. There are different types of rollers in use.

   i) In *Iron rollers* cast iron rings of 0.6 m in diameter are fixed to an axle and provided with a hoped frame for hitching with power unit. Surface of the roller may be plain or fluted or ribbed. The fluted and ribbed rollers are more efficient in breaking clods. The plain rollers are used for compacting the soil surface.

   ii) *Stone rollers* are commonly used for threshing grains. They are made up of cylindrical stone with 0.75 m length and 0.4 m diameter.

   iii) *Sheep foot roller* is the latest implement developed by TNAU, to create partial compaction in paddy fields in light soils. It has a cylindrical drum with projections on the surface like sheep foot.

Sowing Implements

a) ***Country seed drill/'Gorru'***: It consists of a horizontal beam on which a number of tynes are fixed at suitable distances. The tyne is like the body of the common wooden plough, but is much smaller. It has a vertical hole, a little above the point of penetration into the soil. Seeds are released from the above placed seed hoppers steadily few seeds at a time. The base of the hopper has as many holes as there are tynes in the gorru and narrow bamboo or metal tubes connect the hopper and the tynes. This enables the seeds released in the hopper being dropped in the furrows opened by the tynes. The hopper and the seed tubes are held in position with thin ropes.

b) ***Mechanical seed drills***: It is of both bullock and tractor drawn.

   i) ***Bullock drawn seed drill / TNAU improved planter***: A medium size five tyned cup feed seed drill suitable for heavy size bullocks, a small three tyned cup feed seed drill called as *Kovai seed drill* are suitable for small pair of bullocks. These drills are suitable for sowing seeds of groundnut, maize, sorghum, cotton, bengal gram and pulses. It covers on ha per day and cost Rs. 3,500/-.

ii) ***Tractor drawn seed drill***: Both simultaneous formation of 1.5 m wide beds and sowing in the bed is possible using this drill. The implement consists of a pair of furrowers made of sheet metal with suitable hitching arrangements to the three point linkage of the tractor. Over the frame work of these furrowers, 7 numbers of hoppers with metering mechanisms have been mounted. This implement simultaneously sows in seven rows in the broad bed. It covers an area of 4 ha/day and saves 25% of sowing cost.

iii) ***Paddy drum seeder for wet land: (Drum*** seeder for direct sowing of paddy) A manually pulled, paddy seeder has been developed at TNAU for sowing pre-germinated paddy seeds in rows directly in well puddled and levelled soil. It requires 2 labourers and covers 0.4 ha/day. Using this seeder green manures (*Sesbania sp.*) can also be sown as intercrop in between rice rows. Cost of this seeder is Rs.3000/-.

iv) **Paddy-green manure drum seeder** Souring of paddy and green manure can be done manually together.

Implements for Wetlands

Implements for wetlands: Under wetland system the land is prepared by puddling for planting wet rice. **Puddling** means mechanical manipulation of saturated soils with standing water in the field. Actually the structure of the soil is destroyed under puddling. The optimum depth of puddling is about 10 cm in the clay and clay-loam types of soils. Good puddling or neatly ploughed means the soil should be soft, uniformly levelled without weeds or stubbles and with minimum percolation.

Objectives or Advantages of puddling: Puddling is done

i) To obtain a soft seedbed for the seedling to establish faster,

ii) To minimize percolation of water so that water can stagnate in the field,

iii) To minimize leaching loss of nutrients and thereby increase the availability of plant nutrients,

iv) To facilitated better availability of nutrient by achieving reduced soil condition,

v) To incorporate the weeds and stubble in to the soils and

vi) To minimize the weed problems

The implements used for puddling the wet soils are as follows:

i) Country plough, ii) Bose plough, iii) Wet land puddler, iv) Cagewheel, v) Sheep foot roller and vi) Helical bladed puddler

i), ii) and v) were discussed already in this chapter.

iii) ***Wetland puddler***: It consists of three angular bladed cast iron huds rigidly fixed to a hallow horizontal pipe and is rotated when dragged by a pair of bullocks. This implement is proved to be an economic, labour-saving and an effective dual purpose implement useful for puddling and trampling green leaf manure in the puddle field. When used for trampling the vegetative matter is cut and buried in the soil. It covers an area of 0.8 ha/day.

iv) ***Cage wheel***: It is used for puddling in medium and heavy clayey soils in wet lands for paddy cultivation. The cage wheels are attached in place of pneumatic wheels in power tiller and tractor. The cage wheels perform well in all the fields except in fields with clay and silt content of the soil was more than 56%. It saves cost, time and brings more uniformity and thoroughness in the puddle than country ploughing. Cage wheel attached to power tiller covers an area of 0.44 ha/day. The average depth of puddle is 23 cm.

v) ***Helical bladed puddler***: It is used to puddle the wet land soil after initial ploughing with country plough or melur plough. It is a bullock-drawn implement. Five numbers of helical blades made of mild steel are fixed in a skewed shape and mounted on a wooden frame having wooden bearing such that the blades can rotate freely. A handle and pole shaft are provided. Due to the helical shape of the blade, there will be continuous contact between the blades and the soil which gives uniform load on the neck of the bullocks. After ploughing the land with country plough, the implement can be used to puddle the soil up to a depth of 10 cm. The helical geometry facilitates better churning and slicing of the soil requied for puddling. It covers 0.6 ha/day. (fig. 18)

Questions

Fill the blanks

1. Tillage done at any time with some special purpose is called __________
2. Tillage done to break the subsoil hardpan is called __________

3. Tillage that aims at reducing tillage operations to the minimum necessity for ensuring a good seedbed is termed as ________
4. Buck scraper is an implement used for ________
5. Mechanical manipulation of saturated soil with standing water in the field is ________

Choose the correct answer

6. Mechanical manipulation of the soil to make it suitable for crop production is
   a. Tilth b. Tillage
   c. Preparatory tillage d. Primary tillage
7. The tillage operation done after the harvest of crop to bring the land under cultivation is
   a. Primary tillage b. Secondary tillage
   c. Special tillage d. Off season tillage
8. Tillage operations done for raising the crops in the same season or at the onset of the crop season are called as on season tillage
   a. On season tillage. b. Secondary tillage
   c. Special tillage d. Off season tillage
9. Tillage done to cut open/break the subsoil hard pan or plough pan is
   a. Subsoil tillage b. Wet tillage
   c. Inter tillage d. Strip tillage
10. Desirable ploughing depth is
    a. 15 to 30 cm b. 10-15 cm
    c. 40-60 cm d. 60-70 cm

Answer the following

1. Objectives of tillage
2. Special purpose tillage
3. Zero tillage (No tillage)
4. Advantages of puddling
5. Time of ploughing (Based on moisture status and soil type)

□□□

10

Seeds and Sowing

Seed is a fertilized ripened ovule consisting of three main parts namely seed coat, endosperm and embryo which in due course gives rise to a new plant. Seed is the nucleus of life. Endosperm is the storage organ for food substance that nourishes the embryo during its development. Seed coat is the outer cover that protects or shields the embryo and endosperm.

Importance

Plants reproduce sexually by seeds and asexually by vegetative parts. Grains which are used for multiplication are called seeds while those used for human or animal consumption are called grains. Good stalks of planting materials are basic to profitable crop production. The seed or planting material largely determines the quality and quantity of the produce. A good seed or stalk of planting material is genetically satisfactory and true to type, fully developed and free from contamination, deformities, diseases and pests.

Characteristics of A Good Quality Seed

A good quality seed should possess the following characteristics.

1. Seed must be true to its type i.e., genetically pure, free from admixtures and should belong to the proper variety or strain of the crop and their duration should be according to agro climate and cropping system of the locality.
2. Seed should be pure, viable, vigorous and have high yielding potential.
3. Seed should be free from seed borne diseases and pest infection.
4. Seed should be clean; free from weed seeds or any inert materials.
5. Seed should be in whole and not broken or damaged; crushed or peeled off; half filled and half rotten.

6. Seed should meet the prescribed uniform size and weight of a crop
7. Seed should be as fresh as possible or of the proper age
8. Seed should contain optimum amount of moisture (8 to 12%)
9. Seed should have high germination percentage (more than 80%)
10. Seed should germinate rapidly and uniformly when sown

Advantages of using good quality seeds or planting stalks

i) Reduced cost of cleaning, standardization and disinfection.

ii) Uniform germination thus avoiding replanting, gap filling.

iii) Vigorous seedling growth which reduces weed and disease damages.

iv) Uniform growth stages, maturity and products.

v) Maintain good quality under storage conditions.

vi) Reduced cost of seeds.

Seed Rate

The required number of plants/unit area is decided by calculating the seed rate. The seed rate depends on spacing or plant population, test weight, germination percentage. The formula is as follows.

$$\text{Seed rate (kg/ha)} = \frac{\text{Plant population (per ha)} \times \text{No. of seeds/hill} \times \text{Test weight (g)}}{1000 \times 1000 \times \text{Germination percentage (\%)}} \times 100$$

Seed Treatment

Seed treatment is a process of application either by mixing or by coating or by soaking in solutions of chemicals or protectants (with fungicidal, insecticidal, bactericidal, nematicidal or biopesticidal properties), nutrients, hormones or growth regulators or subjected to a process wetting and drying or subjected to reduce, control or repel disease organisms, insects or other pests which attack seeds or seedlings growing there from. Seed treatment also includes control of pests when the seed is in storage and after it has been sown/planted.

Objectives

i) To protect from seed borne diseases and pests.

ii) To protect from or repel birds and rodents.

iii) To supply plant nutrients.

iv) To inoculate micro organisms.

v) To supply growth regulators.

vi) To supply selective herbicides.

vii) To break seed dormancy.

viii) To induce drought tolerance.

ix) To induce higher germination percentage, early emergence.

x) To obtain polyploids (genetic variation) by treating with x-rays, gamma rays and colchicines.

xi) To facilitate mechanized sowing.

Methods of Seed Treatment

1. *Dry treatment*: Mixing of seed with powder form of pesticides / nutrients.
2. *Wet treatment*: Soaking of seed in pesticide / nutrient solutions
3. *Slurry treatment*: Dipping of seeds / seedlings in slurry. Example – rice seedlings are dipped in phosphate slurry.
4. *Pelleting*: It is the coating of solid materials in sufficient quantities to make the seeds larger, heavier and to appear uniform in size for sowing with seed drills. Pelleting with pesticides as a protectant against soil organisms, soil pests and as a repellant against birds and rodents.

Sowing

Sowing is the placing of a specific quantity of seeds in the soil for germination and growth while planting is the placing of plant propagules (may be seedlings, cuttings, rhizomes, clones, tubers etc.) in the soil to grow as plants.

Sowing Methods

1. Broadcasting
2. Dibbling
3. Sowing behind the country plough (manual and mechanical drilling)

4. Seed drilling
5. Nursery transplanting

1. Broadcasting

- Broadcasting is otherwise called as random sowing.
- Literally means 'scattering the seeds'.
- Broadcasting is done for many crops.
- Broadcasting is mostly followed for small sized to medium sized crops.
- This is the largest method of sowing followed in India, since; it is the easiest and cheapest and requires minimum labours.
- To have optimum plant population in unit area certain rules should be followed:
  - Only a skilled person should broadcast the seeds for uniform scattering.
  - The ploughed field should be in a perfect condition to trigger germination.
- The seeds are broadcasted in a narrow strip and the sowing is completed strip by strip.
- To ensure a good and uniform population, it is better to broadcast on either direction.
- This is called criss-cross sowing.
- If the seed is too small, it is mixed with sand to make a bulky one and for easy handling. Ex. Sesame seeds are mixed with sand at 1:15 or 1:10 ratio and sown.
- In certain cases the person sowing will be beating the seeds against the basket for uniform scattering. Ex. Sorghum, pearl millet.
- After broadcasting, the seeds are covered gently either using a country plough with a very shallow ploughing or some wooden planks (boards/levellers) are used to cover the surface.
- In some cases, tree twigs or shrub branches are used.
- If the seeds are large, levellers collect the seeds and leave in the other side.
- Comb harrow is the best used one.

Disadvantages

- All the seeds broadcasted do not have contact with the soil. 100% germination is not possible.
- Enhanced seed rate is required.
- Seeds cannot be placed in desired depth. Desired depth ensures perfect anchorage. Lodging (falling down) is common in broadcasting.

2. Dibbling

- This is actually called line sowing. Inserting a seed through a hole at a desired depth and covering the hole.
- Dibbling is practiced on plain surface and ridges and furrows or beds and channels.
- This type of sowing is practiced only under suitable soil condition.
- Rice fallow cotton is dibbled on a plain surface.
- The seeds are dibbled at 2/3rd from top or 1/3rd at bottom of the ridge.
- Before sowing, furrows are opened and fertilizers are applied above which seeds are sown.
- The seeds do not have contact with the fertilizers.
- This is done for wider spaced crops and medium to large sized seeds. Ex. Sorghum, maize, sunflower, cotton are dibbled on ridges and furrows.
- Both beds and channels; and ridges and furrows come under line sowing.
- While earthing up, the plant occupies middle of the ridge.
- Earthing up is essential for proper anchorage of the root system.

Advantages of line sowing are,

i) uniform population,

ii) better germination and

iii) reduced seed rate.

S. No.	Dibbling (Line sowing)	Broadcasting (Random sowing)
1.	Costlier	Cheaper
2.	Takes considerable time	Quickest and time saving
3.	Fixed seed rate	Higher seed rate
4.	Mechanization is possible, e.g. weeding	Not possible
5.	Uniform utilization of resources (land, water, light, nutrient, etc.)	Resource utilization is un-uniform

3. Sowing behind the plough

- Sowing behind the plough is done by manual or mechanical means.
- Seeds are dropped in the furrows opened by the plough and the same is closed or covered when the next furrow is opened.
- The seeds are sown at uniform distance.
- Manual method is a laborious and time consuming process.
- Seeds like redgram, cowpea and groundnut are sown behind the country plough.
- Major crop sown is groundnut.

4. Drill sowing (or) Drilling

- Drilling is the practice of dropping seeds in a definite depth covered with soil and compacted.
- In this method, sowing implements are used for placing the seeds into the soil.
- Both animal drawn Gorrus and power operated (seed drills) implements are available.
- Seeds are drilled continuously or at regular intervals in rows.
- In this method, depth of sowing can be maintained and fertilizer can also be applied simultaneously.
- It is possible to take up sowing of inter crops also.
- It requires more time, energy and cost, but maintains uniform population per unit area.
- Seeds are placed at uniform depth, covered and compacted.
- Seeds are sown by mechanical means by Gorrus – seed drill.

- A seed drill has a plough share and hopper.
- Seeds are placed on hopper.

Advantages

i) The seeds are placed at desired depth covered by iron planks,

ii) except very small and very large seeds most of the seeds can be sown,

5. Transplanting

This method of planting has two components, a. nursery and b. transplanting. In nursery, young seedlings are protected more effectively in a short period and in a smaller area. Management is easy and economical.

- Age of seedlings is 1/4th of the total duration of the crop.
- If the total duration is 16 weeks, four week period (1 month) is under nursery beds.
- Nursery age is not very rigid, e.g., thumb rule – 3 months crop – nursery duration 3 weeks, minimum 4 months – 4 weeks minimum period; 5 months – 5 weeks.
- After the nursery period, seedlings are pulled out and transplanted.
- This is done on the main field after thorough field preparation or optimum tilth.
- The seedlings are dibbled in lines or in random.
- Closer spaced crops are mostly raised in random method even after nursery, Ex. Rice and finger millet.
- For vegetables, desired spacing is required during transplanting.
- Transplanting shock is a period after transplanting, the seedlings show no growth.
- This is mostly due to the change in the environment between root and the soil.
- The newly planted seedlings should adjust with new environment.
- It is for a period of 5-7 days depending upon season, crop, variety, etc.
- At higher temperature, dehydration is possible and leaves dried out.
- Area required for nursery normally is 1/10th of the total area.

Advantages

- Can ensure optimum plant population
- Sowing of main field duration, i.e., management in the main field is reduced
- Crop intensification is possible under transplanting

Disadvantages

- Nursery raising is expensive
- Transplanting is another laborious and expensive method

Germination

- Germination is a protrusion of radicle or seedling emergence.
- Germination results in the rupture of the seed coat and emergence of seedling from embryonic axis.

Factors affecting seed germination

1. **Soil:** Soil type, texture, structure and microorganisms greatly influence the seed germination.
2. **Moisture:** When the seeds do not get required moisture in the soil, the viability is lost. When the moisture is excess after germination, it will lead to rotting of the sprouts.
3. **Temperature:** When it is above and below the optimum temperature, the germination rate will be affected. The optimum temperature is that one gives the highest germination percentage in the shortest period of time.

	Minimum ^{0}C	Optimum ^{0}C	Maximum ^{0}C
Maize	8-10	20-25	30-35
Rice	10-12	20-27	30-32
Wheat	3-5	15-31	33

4. **Light:** The most effective wavelength for promoting germination is red (662 nm) and 730 nm inhibits germination.

5. **Soil condition**

a) Tilth is the most important soil factor influences on germination of seed. Small seeds require fine tilth whereas, moderate and larger seeds requires medium and coarse tilth soils, respectively.

b) Depth of sowing:

- The seeds should be placed at optimum depth.
- When the seeds are placed at deeper layers they have to spend more energy for germination.
- When it is placed on soil surface, it will be taken away by birds/ worked away.
- The thumb rule is to sow seeds to a depth of approximately 3 to 4 times diameter of the seed.
- The optimum depth of sowing for most of the field crops ranged between 3 and 5 cm depth.
- The seeds sown should be protected from rodents or birds before germination by employing labourers to scare the birds at least for three days after sowing.

6. **Exogenous chemicals**: Some chemicals induce or favour quick and rapid germination, they are :

i) Gibberellins stimulate germination in protoplasmic seeds.

ii) Hydrogen peroxide (H_2O_2) is used for legumes, tomato and barley.

iii) Ethylene (C_2H_4) is used for stimulating groundnut germination.

Factors involved in sowing management

This can be classified into 2 broad groups.

1. **Mechanical factors:** Such as depth of sowing, emergence habit, seed size and weight, seed bed texture, seed - soil contact, seedbed fertility, soil moisture etc.

2. **Biological factors:** Like companion crops, competition for light, soil microorganisms etc.

1. Mechanical factors

i) ***Seed size and weight***: Heavy and bold seeds produce vigorous seedlings. Application of fertilizer to bold seed tends to encourage the seedlings than the seedlings from small seeds.

ii) *Depth of sowing*: Optimum depth of sowing ranges from 2.5 - 3cm. Depth of sowing depends on seed size and availability of soil moisture. Deeper sowing delays field emergence and thus delays crop duration. Deeper sowing sometimes ensures crop survival under adverse weather and soil conditions mostly in dry lands.

iii) *Emergence habit*: Hypogeal seedlings may emerge from a relatively deeper layer than epigeal seedlings of similar seed size.

iv) *Seed bed texture*: Soil texture should minimize crust formation and maximize aeration which in turn influences the gases, temperature and water content of the soil. Very fine soil may not maintain adequate temperature and water holding capacity.

v) *Seeds - Soil contact*: Seeds require close contact with soil particles to ensure that water can be absorbed readily. A tilled soil makes the contact easier. Forming the soil around the seed (Broadcasted seeds) after sowing improves the soil - seed contact.

vi) *Seedbed fertility*: Tillering crops like rice, ragi, bajra etc. should be sown thinly on fertile soils and more densely on poor soils. Similarly high seed rate are used on poor soil for non-tillering crops. Although higher the seed rate greater the yield under conditions of low soil fertility, in some cases such as cotton, a lower seed rate gives better result than a higher seed rate.

vii) *Soil moisture*: Excess moisture in soil retards germination and induce rotting and damping off disease except in swamp (deep water) rice. Adjustment in depth is made according to moisture conditions, i.e., deeper sowing on dry soils and shallow sowing on wet soils. Sowing on ridges is usually recommended on poorly drained soils.

2. Biological Factors

i) *Companion crop*: Is usually sown early to suppress weed growth and control soil erosion. In cassava + maize/yam cropping, cassava is planted later in yam or maize to minimize the effect of competition for light. In mixed cropping all the crops are sown at the same time.

ii) *Competition of light*: In mixed stands, optimum spacing for each crop minimizes the competition of light.

iii) *Soil micro organisms*: The micro organisms present in the soil should favour seed germination and should not posses any harmful effect on seeds / emerging seedlings.

Questions

Fill the blanks

1. Broadcasting is otherwise called as ____________.
2. Mixing of seeds with powder form of pesticides/ nutrients is termed as ____________.
3. Inserting a seed through a hole at a desired depth and covering the hole is called ____________.
4. The yearly sequence and spatial arrangement of crops or of crops and fallow on a given area is ____________.
5. The area required for nursery is normally ____________ of the total area.

Choose the correct answer

6. Dipping of seeds / seedlings in slurry is called as

   a. Slurry treatment b. Wet treatment

   c. Dry treatment d. Hot water treatment

7. The practice of dropping seeds in a definite depth covered with soil and compacted

   a. Drilling b. Broadcasting

   c. Dibbling d. Transplanting

8. The largest method of sowing followed in India is

   a. Drilling b. Broadcasting

   c. Dibbling d. Transplanting

9. While earthing up, the plant occupies

   a. Middle of ridge b. Sides of ridge

   c. Side of furrows d. Middle of furrows

10. The optimum amount of moisture to be present in the seeds is

   a. 8-12 % b. 4-6 %

   c. 14-16 % d. 20-22 %

Answer the following

1. Sowing methods
2. Factors affecting seed germination
3. Objectives of seed treatment
4. Wet seed treatment
5. Seed pelleting.

11

Plant Population and Crop Geometry

Crop Stand Establishment

Good crop establishment is one of the most important features in better crop production. The better crop establishment is in turn expressed as optimum plant population in fields.

Plant population is the number of plants per unit area in a cropped field. It indicates the size of the area available for individual plant.

Optimum Plant Population

It is the number of plants required to produce maximum output or biomass per unit area. Any increase beyond this stage results in either no increase or reduction in biomass.

Crop geometry is the pattern of distribution of plant over the ground or the shape of the area available to the individual plant, in a crop field.

Importance of Plant Population

- Yield of any crop depends on final plant population.
- The plant population depends on germination percentage and survival rate in the field.
- Under rain fed conditions, high plant population will deplete the soil moisture before maturity, where as low plant population will leave the soil moisture unutilized.
- When soil moisture and nutrients are not limited high plant population is necessary to utilize the other growth factors like solar radiation efficiently.
- Under low plant population, individual plant yield will be more due to wide spacing.

- Under high plant population, individual plant yield will be low due to narrow spacing leading to competition between plants.
- Yield per plant decreases gradually as plant population per unit area is increased, but yield per unit area increases up to certain level of population. That level of plant population is called as optimum population.
- So, to get maximum yield per unit area, optimum plant population is necessary. So the optimum plant population for each crop should be identified.

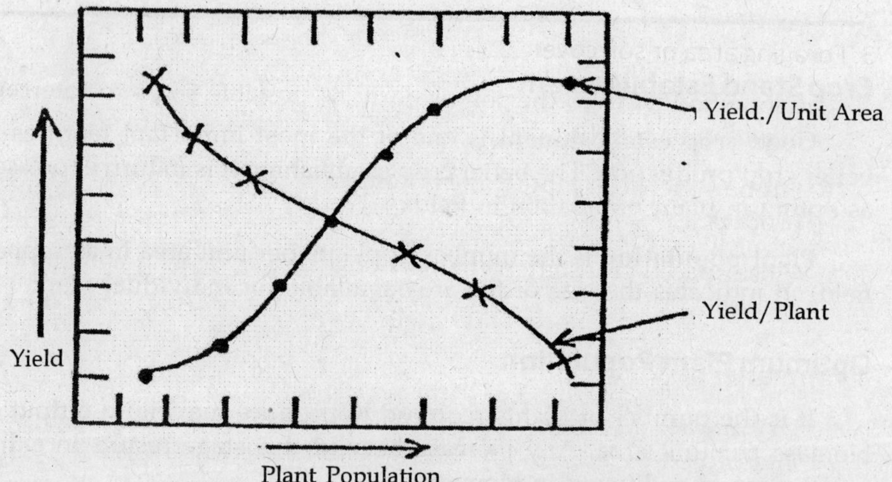

Fig. 19. Influence of Plant Population on the Yield

Factors Affecting Plant Population

A. Genetic factors

1. Size of the plant

- The volume occupied by the plant at the time of flowering decides the spacing of the crop.
- Plants of red gram, cotton, sugarcane etc. occupy larger volume of space in the field compared to rice and wheat.
- Even the varieties of the same crop differ in size of the plant.

2. Elasticity of the plant

- Variation in size of the plant between minimum size of the plant that can produce some economic yield to the maximum size of the plant that can reach under unlimited space and resources is the elasticity of the plant.

- The optimum plant population range is high in indeterminate plants. Ex. Redgram – 55,000 to1,33,000 plants/ha.
- The elasticity is due to tillering and branching habit of the plants.
- For determinate plants like pearl millet, sorghum elasticity range is less.
- For indeterminate plants like cotton and Redgram, more branches will be produced by the crop.

3. Foraging area or soil cover

- Crop should cover the soil as early as possible so as to intercept maximum sunlight.
- Higher the intercepted radiation more will be the dry matter produced.
- Close spaced crops intercept more solar radiation than wide spaced crops.

4. Dry matter partitioning

- Dry matter production is related to amount of solar radiation intercepted by the canopy which depends on plant density.
- As the plant density increases, the canopy expands more rapidly, more radiation is intercepted and more dry matter is produced.

5. Crop and variety

Depending on the crops and varieties, the plant population varies.

Rice	Short duration	6,66,666 plants/ha (15 cm x 10 cm)
	Medium	5,00,000 plants/ha (20 cm x 10 cm)
	Long	3,33,000 plants/ha (20 cm x 15 cm)
Cotton	Medium	55,555 plants/ha (60 cm x 30 cm)
	Long	44,444 plants/ha (75 cm x 30 cm)
	Hybrids	18,518 plants/ha (120 cm x 45 cm)
Maize	Varieties	83,333 plants/ha (60 x 20 cm)
	Hybrids	47,620 plants/ha (60 x 35 cm)

B. Environmental factors

1. Time of sowing

- The crop is subjected to various weather conditions when sown at different periods.
- Among weather factors, day length and temperature influence much on the plant population. As low temperature retards growth, high plant population is required to cover the soil.

2. Rainfall / irrigation

1. Plant population has to be less under rainfed than irrigated condition.
2. Under more plant densities, more water is lost through transpiration.
3. Under adequate rainfall/irrigation, high plant population is recommended.

3. Fertilizer application

1. Higher plant population is necessary to fully utilize higher level of nutrients in the soil to realize higher yield.
2. Nutrient uptake increases at optimum plant population.
3. High population under low fertility soils leads to nutrient deficiency symptoms leading low yield.

4. Seed rate

- Quantity of seed sown/unit area, viability and establishment rate decides the plant population. Under broadcasting the seed rate is higher when compared with line sowing/transplanting, Ex. Rice.

  Direct sowing- 100 kg/ha; Line sowing - 60 kg/ha; Transplanting- 40 kg/ha.

Plant Geometry

The arrangement of the plants in different rows and columns in an area to utilize the natural resources efficiently is called crop geometry. It is otherwise area occupied by a single plant Ex. Rice - 20 cm x 15 cm. This is very essential to utilize the resources like light, water, nutrient and space. Different geometries are available for crop production

Different crop geometries are available for crop production

1. Random plant geometry

Random plant geometry results due to broadcasting method of sowing and no equal space is maintained. Resources are either under utilized or over exploited.

2. Square plant geometry

The plants are sown at equal distances on either side. Mostly perennial crops, tree crops follow square method of cultivation. Ex. Coconut – 7.5 x 7.5 m; banana – 1.8 x 1.8 m. But, due to scientific invention, the square geometry concept is expanded to close spaced field crops like rice too.

Advantages

Light is uniformly available, movement of wind is not blocked and mechanization can be possible.

3. Rectangular method of sowing

There are rows and columns, the row spacing are wider than the spacing between plants.

The different types exist in rectangular method are,

a) *Solid row:* Each row will have no proper spacing between the plants. This is followed only for annual crops which have tillering pattern. There is definite row arrangement but no column arrangement, Ex. Wheat.

b) *Paired row arrangement*: It is also a rectangular arrangement. It a crop requires 60 cm x 30 cm spacing and if paired row is to be adopted the spacing is altered to 90 cm instead of 60 cm in order to accommodate an intercrop. The base population is kept constant.

c) *Skip row*: A row of planting is skipped and hence there is a reduction in population. This reduction is compensated by planting an intercrop; practiced in rainfed or dryland agriculture.

d) *Triangular method of planting:* It is recommended for wide spaced crops like coconut, mango, etc. The number of plants per unit area is more in this system.

Inter Cultivation

Cultivation practices taken up after sowing of crop is called inter-cultivation. It is otherwise called as after cultivation operation. There are three important after cultivation processes viz., Thinning and gap filling, weeding and hoeing and earthing up.

1. Thinning and Gap filling

The objective of thinning and gap filling process is to maintain optimum plant population. Thinning is the removal of excess plants leaving healthy seedlings. Gap filling is done to fill the gaps by sowing of seeds or transplanting of seedlings in gap where early sown seed had not germinated. It is a simultaneous process. Gap filling is done reasonably early so that plants come to maturity along with other plants. Time may vary with duration of crops. For example, in sugarcane it may be done even 30 days after planting. But in short duration crops like maize, sorghum, rice etc., it should be done within about 10 to 15 days.

Normally, these are practiced a week after sowing to a maximum of 15 days. In dryland agriculture, gap filling is done first. Seeds are dibbled after 7 days of sowing. Thinning is done after gap filling; in order to avoid drought. It is a management strategy to remove a portion of plant population to mitigate stress is referred to as mid season correction.

2. Weeding and Hoeing

Weeding is removal of unwanted plants. Weeding and hoeing is a simultaneous operation. Hoeing is disturbing the top soil by small hand tools and helps in aerating the soil.

3. Earthing up

It is a dislocation of soil from one side of a ridge and to be placed nearer the cropped side. It is carried out in wide spaced and deep rooted crops. It is done around 6-8 weeks after sowing / planting in sugarcane, tapioca, banana, etc.

4. Other inter cultivation practices

Harrowing

Stirring or scraping the surface soil in inter and intra row spacing of the crop using tools or implements.

Roguing

Removal of plants of a variety admixed with other variety of same crop. Ex. In IR 50 rice field, the other rice varieties are rogue. It is practiced in seed production to maintain purity.

Topping

Removal of terminal buds. It is done to stimulate auxillary growth. Practiced in cotton and tobacco.

Propping

Provision of support to the crop is called propping. Practiced in sugarcane commonly. It is done to prevent lodging of the crop. Cane stalks from adjacent rows are brought together and tied with their own trash and old leaves.

De-trashing

Removing of older leaves from the sugarcane crop.

De-suckering

Removal of axillary buds and branches which are considered non essential for crop production and which removes plant nutrients considerably are called suckers. Ex. Tobacco.

Questions

Fill the blanks

1. The number of plants required to produce maximum output or biomass per unit area is termed as ______________.
2. The pattern of distribution of plant over the ground or the shape of the area available to the individual plant, in a crop field is called ______________.
3. In ________ plant geometry, plants are sown at equal distances on either side
4. Yield of any crop depends on final ______________.
5. Under high plant population, individual plant yield will be ________ due to narrow spacing leading to competition between plants.

Choose the correct answer

6. Under low plant population, individual plant yield will be

   a. More b. Less

   c. Optimum d. Nil

7. Removal of axillary buds and branches which are considered non essential for crop production

   a. De-suckering b. Topping

   c. Roguing d. Propping

8. The removal of excess plants leaving healthy seedlings is called

   a. Gap filling b. Thinning

   c. Hoeing d. Earthing up

9. The optimum plant population range is high in

   a. Indeterminate plants b. Determinate plants

   c. Tall crops d. Short crops

10. The seed requirement of crops is high under

    a. Dibbling b. Broadcasting

    c. Drilling c. Transplanting

Answer the following

1. Plant geometry
2. Gap filling
3. Importance of plant population Smother crops
4. Earthing up
5. Paired row planting.

12

Weeds

Weeds are the plants, which grow where they are not wanted. Jethro Tull (1731) coined the word weed. Weeds can also be referred to as plants out of place.

Definition

Weeds are unwanted and undesirable plant that interfere with utilization of land and water resources and thus, adversely affect crop production and human welfare.

In organic agriculture weeds are treated as childrens of the soil.

In the world 30,000 species of weeds have been listed, of which nearly 18,000 cause serious damage to agricultural production. Eighteen weeds are considered as the most serious in the world and about twenty six species have been listed as principal weeds in crop fields of India, and are listed in *annexure V*

Characteristics of Weeds

1. Weeds are more competitive than the cultivated crops.
2. They are capable of thriving under stress condition.
3. They produce enormous seeds.
4. The weeds seeds are easily germinated; they undergo pollination very easily; they have the capacity to withstand adverse situations.

Harmful Effects of Weeds

1. Weeds compete with main crop for space, light, moisture and soil nutrients thus, causing reduction in yield.
2. Affect quality of farm produce, livestock products such as milk and skin.

3. Act as alternate host for pests and pathogens.
4. Cause health problems to human beings. Ex. *Parthenium* causes allergy.
5. Increase cost of cultivation due to weeding problem.
6. Aquatic weeds transpire large quantity of water, obstruct flow of water.
7. Reduce the land value (if *Cynodon* and *Parthenium* are present in the land).
8. Some weeds are poisonous to livestock.

Beneficial Effects of Weeds

1. Weeds are indirectly responsible for crop cultivation, but for them cultivated crops may not receive much attention
2. Manure: When weeds are ploughed in, they add to the soil plenty of humus. Excellent compost can be made out of many weed plants. e.g. *Calotropis gigantea, Croton sparsiflorus* (Syn: *C. bonflandianum*) and *Tephrosia purpurea* are used as green leaf manure for rice.
3. As human food: Weeds serve as human food e.g. *Amaranthus viridis* and *Digera arvensis* used as greens.
4. As fodder: Most weeds are eaten by cattle and weeds like *Rynchosia aurea*, and *Clitoria ternatea* are very good fodder legumes.
5. Weed as fuel: *Prosopis juliflora* very invasive in nature and notorious tree weed commonly used as fire wood. People make charcoal out of it and marketed.
6. Weed as soil binders: *Panicum repens* is an excellent soil binder; keeps bunds in position and prevents soil erosion.
7. Weed as medicine: Many weeds have great therapeutic properties and used as medicine.Eg. *Phyllanthus niruri* – Jaundice; *Eclipta alba* – Scorpion sting; *Centella asiatica* – Improves memory; *Cyanodon dactylon* – Asthma, piles; *Cyprus rotundus* – Stimulates milk secretion.
8. Weed as mats and screens: Stems of *Cyprus pangorei* and *Cyprus corymbosus* are used for mat making, while *Typha angustata* is used for making screens.
9. Weed as indicators: Weeds are useful as indicators of good and bad soils. *E. colonum* occurs in rich soils while *Cymbopogon* denotes poor light soil and sedges are found in ill-drained soils.

Crop Weed Competition

Competition for nutrients

The nutrient concentration of weeds far exceeds the associated crop. Nutrient taken up by maize and its associated weeds show that weeds contain higher concentration of nutrients.

Competition for light

Weeds deplete photosynthetically active radiation resulting in reduction of photosynthesis and shortening the life of lower leaves. Due to shading, crop plants may have thin, pale green leaves with reduced number of vascular bundles and meristem cells.

Competition for water

The amount of water required to produce a unit amount of drymatter is known as transpiration ratio. The transpiration ratio of weeds is higher compared to crops. The transpiration ratio of *Cyanodon dactylon* is around 813 compared to 450 for pearl millet and 430 for sorghum. This indicates that more water is required to grow one unit of weeds than crops.

Competition for CO_2

Competition for CO_2 may occur when the weed infestation is high. Since most of the weeds are C_4 plants, they can deplete very low concentration of CO_2 compared to crops.

Allelopathic effect

Most of the plants excrete certain chemicals into soil which inhibit germination and growth of other plants in their vicinity and this phenomenon is called as allelopathy. The degree of allelopathic effect depends on the plant species. Important weed species that show allelopathic effect are *Agropyron repens, Sorgum halepense, Lantana camara, Cyprus rotundus* etc.

Critical period of weed competition

It is defined as the shortest time span during the crop growth when weeding results in the highest economic returns. The crop yield level obtained by weeding during this period is almost similar to that obtained by full season weed free conditions. The critical period of weed competition is around 30 days for most of the crops.

Critical period of weed competition for different crops

Crop	Critical period (Days after sowing)	
	From	To
Rice (transplanted)	15	45
Upland rice	Entire period	
Sorghum	15	45
Maize	15	35
Finger millet	25	45
Soybean	15	45
Blackgram	30	45

Methods of Weed Control

A. Mechanical method

Mechanical or physical methods of weed control are being employed ever since man began to grow crops. The mechanical methods include tillage, hoeing, hand weeding, digging, cheeling, sickling, mowing, burning, flooding, mulching, etc.

1. Tillage

Tillage removes weeds from the soil resulting in their death.It may weaken plants through injury of root and stem pruning, reducing their competitiveness or regenerative capacity. Tillage also buries weeds. Tillage operation includes ploughing, discing, harrowing and levelling which is used to promote the germination of weeds through soil turnover and exposure of seeds to sunlight, which can be destroyed effectively later. In case of perennials, both top and underground growth is injured and destroyed by tillage.

2. Hoeing

Hoe has been the most appropriate and widely used weeding tool for centuries. It is a very useful implement to obtain results effectively and cheaply. It supplements the cultivator in row crops. Hoeing is particularly more effective on annuals and biennials as weed growth can be completely destroyed. In case of perennials, it destroyed the top growth with little effect on underground plant parts resulting in re-growth.

3. Hand weeding

It is done by physical removal or pulling out of weeds by hand or removal by implements called Khurpi, which resembles sickle. It is probably the oldest method of controlling weeds and it is still a practical and efficient method of eliminating weeds in cropped and non-cropped lands. It is very effective against annuals, biennials and controls only upper portions of perennials.

4. Digging

Digging is very useful in the case of perennial weeds to remove the underground propagating parts of weeds from the deeper layer of the soil.

5. Cheeling

It is done by hand using a cheel hoe, similar to a spade with a long handle. It cuts and shapes the above ground weed growth.

6. Sickling and mowing

Sickling is also done by hand with the help of sickle to remove the top growth of weeds to prevent seed production and to starve the underground parts. It is popular in slopy areas where only the tall weed growth is sickled leaving the root system to hold the soil in place to prevent soil erosion. Mowing is a machine-operated practice mostly done on roadsides and in lawns.

7. Burning

Burning or fire is often an economical and practical means of controlling weeds. It is used to (a) dispose of vegetation, (b) destroy dry tops of weeds that have matured, (c) kill green weed growth in situations where cultivations and other common methods are impracticable.

8. Flooding

Flooding is successful against weed species sensitive to longer periods of submergence in water. Flooding kills plants by reducing oxygen availability for plant growth. The success of flooding depends upon complete submergence of weeds for longer periods.

Merits of mechanical method

a) Oldest, effective and economical method,

b) Large area can be covered in shorter time,

c) Safe method for environment,

d) Does not involve any skill,

e) Weeding is possible in between plants,

f) Deep rooted weeds can be controlled effectively

Demerits of mechanical method

a) Labour consuming,

b) Possibility of damaging crop and

c) Requires ideal and optimum specific condition.

B. Cultural weed control

Several cultural practices like tillage, planting, fertilizer application, irrigation etc., are employed for creating favourable condition for the crop. These practices if used properly help in controlling weeds. Cultural methods, alone cannot control weeds, but help in reducing weed population. They should, therefore, be used in combination with other methods. In cultural methods, tillage, fertilizer application and irrigation are important. In addition, aspects like selection of variety, time of sowing, cropping system, cleanliness of the farm etc., are also useful in controlling weeds.

1. Field preparation

The field has to be kept weed free. Flowering of weeds should not be allowed. This helps in prevention of build up of weed seed population.

2. Summer tillage

The practice of summer tillage or off-season tillage is one of the effective cultural methods to check the growth of perennial weed population in crop cultivation. Initial tillage before cropping should encourage clod formation. These clods, which have the weed propagules, upon drying desiccate the same. Subsequent tillage operations should break the clods into small units to further expose the shriveled weeds to the hot sun.

3. Mulching

Mulch is a protective covering of material maintained on soil surface. Mulching has smothering effect on weed control by excluding light from the photosynthetic portions of a plant and thus inhibiting the top growth. It is very effective against annual weeds and some perennial weeds like *Cyanodon dactylon*. Mulching is done by dry or green crop residues, plastic sheet or polythene film. To be effective the mulch should be thick enough to prevent light transmission and eliminate photosynthesis

4. Solarisation

This is another method of utilisation of solar energy for the desiccation of weeds. In this method, the soil temperature is further raised by 5-10°C by covering a pre-soaked fallow field with thin transparent plastic sheet. The plastic sheet checks the long wave back radiation from the soil and prevents loss of energy by hindering moisture evaporation.

5. Stale seedbed

A stale seedbed is one where initial one or two flushes of weeds are destroyed before planting of a crop. This is achieved by soaking a well prepared field with either irrigation or rain and allowing the weeds to germinate. At this stage a shallow tillage or non- residual herbicide like paraquat may be used to destroy the dense flush of young weed seedlings. This may be followed immediately by sowing. This technique allows the crop to germinate in almost weed-free environment.

6. Blind tillage

The tillage of the soil after sowing a crop before the crop plants emerge is known as blind tillage. It is extensively employed to minimise weed intensity in drill sowing crops where emergence of crop seedling is hindered by soil crust formed on receipt of rain or irrigation immediately after sowing.

7. Flaming

Use of flamers producing flame or fire through the burning of propane gas.

Merits of Cultural Method

a) Low cost for weed control,

b) Easy to adopt,

c) No residual problem,

d) Technical skill is not involved,
e) No damage to crops,
f) Effective weed control,
g) Crop-weed ecosystem is maintained.

Demerits of Cultural Method

a) Immediate and quick weed control is not possible,
b) Weeds are kept under suppressed condition,
c) Perennial and problematic weeds can not be controlled,
d) Practical difficulty in adoption.

C. Biological weed control

Use of living organism's *viz.*, insects, disease organisms, herbivorous fish, snails or even competitive plants for the control of weeds is called biological control. In biological control method, it is not possible to eradicate weeds but weed population can be reduced. This method is not useful to control all types of weeds. Introduced weeds are best targets for biological control.

Examples of biological control: Water hyacinth is controlled by Hyacinth moth; *Opuntia* is controlled by scale insects; *Parthenium* is by *Zygogramma bicolorata*; Strangler vine (*Morrenia odorata*) controlled by fungus- *Phytophthora palmivora* (Devine) in citrus gardens.

Broherlicodes : Collego from *Collectotrichum* spp. and Devine are bio-herbicides.

D. Chemical weed control

Using chemicals, generally referred as herbicides, for the control of weeds is called chemical weed control. In 1944, discovery of 2,4-D Na salt as a land mark in herbicide usage. Commonly used herbicides are Pendimethalin, Atrazine, Fluchloralin etc.

Merits of chemical weed control

- Herbicide can be recommended for adverse soil and climatic conditions, as manual weeding is highly impossible during monsoon season.
- Herbicide can control weeds even before they emerge from the soil so that crops can germinate and grow in completely weed-free environment at early stages. It is usually not possible with physical weed comfort.

- Weeds, which resemble like crop in vegetative phase, may escape in manual weeding. However, these weeds are controlled by herbicides.
- Herbicide is highly suitable for broadcasted and closely spaced crops.
- Controls the weeds without any injury to the root system of the associated standing crop especially in plantation crops like Tea and Coffee
- Reduces the need for pre planting tillage
- Controls many perennial weed species
- Herbicides control the weed in the field itself or *in-situ* controlling whereas mechanical method may lead to dispersal of weed species through seed
- It is profitable where labour is scarce and expensive
- Suited for minimum tillage concept and highly economical

Demerits

- Pollutes the environment
- Affects the soil microbes if the dose exceeds
- Herbicide causes drift effect to the adjoining field
- It requires certain amount of minimum technical knowledge for calibration
- Leaves residual effects
- Some herbicide is highly costlier
- Suitable herbicides are not available for mixed and inter-cropping system.

Integrated Weed Management

FAO Definition

It is a method whereby all economically, ecologically and toxicologically justifiable methods are employed to keep the harmful organisms below the threshold level of economic damage, keeping in the foreground the conscious employment of natural limiting factors.

Weeds can be controlled by several methods. However, each weed control method has advantages and disadvantages. None is applicable in all crops and weed situations. Continuous use of same method leads to build up of tolerant weeds. It is, therefore, necessary to combine or change the method of weed control. Integrated weed management is a weed population management system that uses all suitable techniques in a combative manner to reduce weed population and maintain them at levels below those causing economic injury. The components of integrated weed management are cultural, physical, biological and chemical methods. By using two or more of these methods, suitable for crop and weed situation, weeds can be controlled effectively and economically.

Questions

Fill the blanks

1. ____________number of weeds are considered as the most serious in the world
2. The weed that causes health problems (allergy) to human beings is ______________
3. ________________ is a method of utilisation of solar energy for the desiccation of weeds.
4. The transpiration ratio of weeds is _________ compared to crops
5. Discovery of __________ is considered as a land mark in herbicide usage.

Choose the correct answer

6. Herbicides that are applied before the emergence of weeds are called

   a. Pre-plant
   b. Pre emergence
   c. Post emergence
   d. Myco herbicides

7. Use of living organism's for the control of weeds is called

   a. Biological control.
   b. Chemical control
   c. Mechanical control
   d. Physical control

8. For most of the crops, the critical period of weed competition is

   a. 10 days
   b. 20 days
   c. 30 days
   d. 50 days

9. The oldest method of controlling weeds is

a. Digging b. Hand weeding

c. Flooding d. Cheeling

10. The herbicide commonly used for pulse crops is

a. 2-4 D b. Pendimethalin

c. Paraquat d. Atrazine

Answer the following

1. Characteristics of weeds
2. Harmful Effects of Weeds
3. Critical period of weed competition
4. Stale seedbed
5. Integrated weed management.

13

Irrigation

Irrigation

Irrigation is defined as the artificial application of water to the soil for the purpose of crop production in supplement to rainfall and ground water contribution.

Time of Irrigation

The time schedule which indicates when irrigation water is to be applied and the quantity of water to be applied at that time is called irrigation schedule. There are several approaches for scheduling irrigation:

1. **Soil moisture depletion approach** based on the available soil moisture in the root zone.
2. **Climatological approach** based on the amount of water lost by evapotranspiration.
3. **Combination approach** based on soil moisture depletion and climatological approaches.
4. **Critical stage approach** which indicates the growth stages at which moisture stress leads to yield loss. These stages are known as critical period or moisture sensitive period. In cereals and millets panicle initiation and flowering stages are moisture sensitive stages. In pulses and oilseeds, flowering and pod development stages are critical stages. In sugarcane active tillering or formative stage is critical and in cotton boll development is the critical stage.

There are also simple techniques for scheduling irrigation like use of tensiometers, feel and appearance method etc.

Irrigation Methods

I. Surface Method of Irrigation

- It is an application method in which water is applied on the surface of the soil.
- It is the oldest (4000 years back) and most common method.
- 90% of world's total irrigated area is under this method.
- In USA also, 66% is by surface irrigation.
- This method is most suitable for low to moderate infiltration rates and leveled lands and having <2-3% slope.
- It is labour intensive.

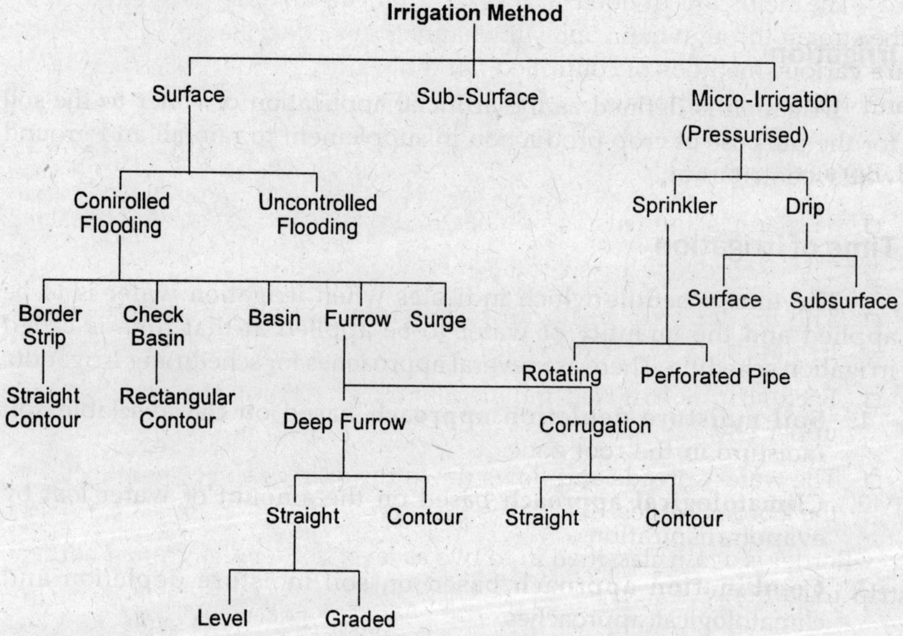

Criteria for selection of surface irrigation

- Topography of land
- Soil type
- Crops to be grown
- Quantity and quality of irrigations water available
- Source and availability of water
- Energy and labour required in conveyance.

The surface irrigation practices are classified as: Uncontrolled flooding and controlled flooding

Uncontrolled flooding

- Primitive and wasteful method.
- Practiced where water is plenty and vast area to be irrigated as in pastures.
- Great loss of water, soil erosion, un-uniform wetting and deep percolation of water are the disadvantages.

Controlled flooding

The fields are divided into several plots and water is guided from the stream through main and sub-channels in a controlled manner. There are various methods of controlled flooding. They are Border, check basin and furrow irrigations

1. Border irrigation

- The land is divided into number of long parallel strips called borders.
- These borders are separated by low ridges.
- The border strip has a uniform gentle slope in the direction of irrigation.
- Each strip is irrigated independently by turning the water in the upper end.
- The water spreads and flows down the strip in a sheet confined by the border ridges.

Border is again classified in to two as level border and graded border strip irrigation.

i. Level border

Level borders ensure uniform distribution of water, avoid erosion from irrigation or rainfall, save labour as one man can guide water to several strips.

Advantages	*Disadvantages*
1. Uniform distribution of water	Require fine grading
2. Avoid erosion	Needs large stream of water
3. Saves labour	

ii. Graded & semi graded

Graded or semigraded borders permit a mild slope to permit fairly fast and uniform spread of water in the border strips.

	Advantages	*Disadvantages*
1.	Light water application is easier	Subjected to erosion
2.	Much land leveling is not required	Requires skill in applying water

Suitability of Border irrigation

Suitable to soils having moderately low to moderately high infiltration rates.

- It is not used in coarse sandy soils that have very high infiltration rates and also in heavy soils having very low infiltration rate.
- Suitable to irrigate all close growing crops like wheat, barley, fodder crops and legumes and not suitable for rice.

Advantages

a) Border ridges can be constructed with simple farm implements like bullock drawn ridger or bund former.

b) Labour requirement in irrigation is reduced as compared to conventional check basin method.

c) Uniform distribution of water and high water application efficiencies are possible.

d) Large irrigation streams can be efficiently used.

e) Adequate surface drainage is provided if outlets are available.

2. Check basin irrigation

- It is the most common surface irrigation method.
- Here, the field is divided into smaller unit areas so that each has a nearly level surface.
- Bunds or ridges are constructed around the area forming basins within which the irrigation water can be controlled.
- The water applied to a desired depth can be retained until it infiltrates into the soil.
- The size of the basin varies from 10 m^2 to 25 m^2 depending upon soil type, topography, stream size and crop.

Suitability of check basin irrigation

- Small gentle and uniform land slopes and soils having moderate to low infiltration rates.
- Adapted to grain and fodder crops in heavy soils and suitable to permeable soils

Advantages

- Check basins are useful when leaching is required to remove salts from the soil profile.
- Rainfall can be conserved and soil erosion is reduced by retaining large part of rain
- High water application and distribution efficiency.

Disadvantages

- The ridges interfere with the movement of implements.
- More area occupied by ridges and field channels.
- The method impedes surface drainage
- Precise land grading and shaping are required
- Labour requirement is higher.
- Not suitable for crops which are sensitive to wet soil conditions around the stem

3. Furrow irrigation

- It is used for row crops.
- The furrows are formed between crop rows.
- The dimension of furrows depend on the crop grown, equipment used and soil type.
- Water is applied by small running streams in furrows between the crop rows.
- Water infiltrates into soil and spreads laterally to wet the area between the furrows.
- In heavy soils, furrows can be used to dispose the excess water.

Suitability

- Method used for wide spaced row crops including vegetables.
- Suitable for maize, sorghum, sugarcane, cotton, tobacco, groundnut, potatoes
- Suitable to most soils except sandy.

Advantages

- Water in furrows contacts only 1/2 to 1/5 of the land surface.
- Labour requirement for land preparation and irrigation is reduced.
- Compared to check basins, there is less wastage of land in field ditches

There are three types of furrow irrigation, they are, all furrow irrigation, alternate furrow irrigation and skip furrow irrigation.

1. **All furrow irrigation:** Water is applied evenly in all the furrows.
2. **Alternate furrow irrigation:** It is a technique for water saving. Water is applied in alternate furrows for eg. During first irrigation if the even numbers of furrows are irrigated, during next irrigation odd furrows are irrigated.
3. **Skip furrow irrigation:** Normally adopted during the period of water scarcity and to accommodate intercrops. Here, a set of furrows are completely skipped out from irrigation permanently. The system ensures water saving of 30-35%.

II. Sub-surface Irrigation

There are four methods as follows :

1. Through porous tile tubes

In this method under ground porous tile tubes are embedded in the sub-surface where root spread is expected. Water passing through these ooze out through the porous holes and wet the root zone of the soil. This is a costly system suitable only for small areas as in lawns.

2. Through underground water table

In this method of subsurface irrigation, water is applied beneath the ground by creating and maintaining an artificial water table at some depth, usually 30-75 cm below the ground surface. Moisture moves upwards towards the land surface through capillary action.

3. Through field trenches

Water is applied through underground field trenches laid 15-30 m apart. Open ditches are preferred because they are relatively cheaper and suitable to all types of soil. The irrigation water should be of good quality to prevent soil salinity.

4. Through subsurface drip irrigation

The drip irrigation components including sub main lines, laterals and dripper are buried under ground at a depth of 30 – 50 cm closer to the root zone so that it does not interfere with crop cultivation operations.

5. Pitcher irrigation

Pitcher Irrigation is an inexpensive small-scale irrigation method. The system consists of burying unglazed clay pots in the soil upto their neck. When the pot is filled with water, the natural pores in the pot's walls allow water to spread laterally in the soil. This creates moist conditions necessary for plant growth. Pitchers are filled as needed, maintaining a continuous supply of water directly to the plant root zone.

Advantages

- Minimum water requirement for raising crops
- Minimum evaporation and deep percolation losses
- No wastage of land
- No interference to movement of farm machinery
- Cultivation operations can be carried out without concern for the irrigation period.

Disadvantages

- Requires a special combination of natural conditions.
- There is danger of water logging
- Possibility of choking of the pipes lay underground.
- High cost.

III. Pressurized or Modern Irrigation Systems

a. Drip irrigation system

- It is also called as trickle irrigation.
- It is one of the latest and modern methods of irrigation.
- It is suitable for water scarcity and salt affected soils.
- Water is applied in the root zone of the crop.

Drip components

- A drip irrigation system consists of a pump or overhead tank, main line, sub-mains, laterals and emitters.
- The mainline delivers water to the sub-mains and the sub-mains into the laterals.
- The emitters which are attached to the laterals distribute water for irrigation.
- The mains, sub-mains are usually made of PVC (poly vinyl chloride) pipes and laterals of Low density poly ethylene (LDPE) tubes. The emitters are also made of PVC material.
- The other components include pressure regulator, filters, valves, water meter, fertilizer application devices like fertigation tanks and ventruy.

Advantages of drip irrigation

- High water use efficiency (~95%, compared to less than 50% in surface)
- Flexibility of wetted area
- Versatile selection of emitters: type, discharge rate, position
- Economy in weed control
- Low interference with cultivation
- Day and night irrigation
- Prevention of leaf wetting
- Energy saving
- Salinity control
- Irrigation at variable topographic conditions.

Disadvantages of drip Irrigation

- High investment
- High level of knowledge for optimal and economical operation
- Susceptibility to mechanical damage
- Large number of emitters
- Long application time
- High level of filtration and other controls
- Clogging problems due to use of saline water necessitating acid flushing.

b. Sprinkler irrigation system

- This is another important modern irrigation techniques followed all over the globe.
- Sprinkler irrigation is application simulating rainfall overhead so called as overhead sprinklers.
- The sprinkler irrigation system conveys water to the field through pipes (aluminium or PVC) under pressure with a system of nozzles.
- This system is designed to distribute the required quantity of water uniformly, which is not possible in surface irrigation.
- Water is applied at a rate less than the infiltration rate of the soil hence the runoff from irrigation is avoided.

Advantages of sprinkler

- Suitable for undulating topography (slopy lands)
- Water saving (35-40%) compared to surface irrigation methods.
- Fertilizers and other chemicals can be applied through irrigation water
- Saving in fertilizers, even distribution and avoids wastage.
- Reduces erosion
- Suitable for coarse textured soils (sandy soils)
- Frost control - protect crops against frost and high temperature
- Drainage problems eliminated
- Saving in land

Disadvantages

- High initial cost
- Efficiency is affected by wind
- Higher evaporation losses in spraying water
- Not suitable for heavy clay soils
- Poor quality water can not be used (Sensitivity of crop to saline water and clogging of nozzles)

Water Requirement of Crop

Water requirement of crop: The quantity of water required by a crop in a given period of time for its normal growth under field conditions i.e.,

Water requirement of crop (WR)	=	Consumptive use + * Application losses + Water needed for special operations (puddling, leaching etc.,)
Irrigation requirement of crop (IR)	=	Water requirement – (Effective RF + Ground water contribution)
Consumptive use (Cu)	=	Evaporation + Transpiration + Water need in building of plant tissue (very small portion)

* Application losses are losses due to deep percolation and seepage in the channels and field.

Water Requirement of crops

Crop	Duration (days)	Water requirement (mm)	No. of irrigation	Irrigation interval (days)	Yield kg/ha	WUE kg/ha mm	Depth of moisture extraction
Rice	135	1,2000	-	-	4,500	3.7	60
Sorghum	110	500	8	9-10	4,500	9.0	120
Maize	110	625	12	5-6	5,000	8.0	120
Finger millet	110	350	10	6-7	4,100	11.7	120
Pearl millet	110	350	10	6-7	4,000	11.4	120
Ground nut	110	510	9	9-10	3,000	5.9	90
Sugarcane	365	2,500	24	7-15	125 t/ha	-	120
Cotton	165	500	10	14-15	3,000	6.0	120
Banana	330	2,000	20	15		-	120

Factors Influencing Water Requirement of Crops

1. Nature of crops

a) Different crops and varieties may differ in their water requirement. Finger millet requires less amount of water (35 cm), whereas rice needs more water (100-120cm) and sugarcane also needs more water (225-250cm).

b) The duration of growing season and stage of crop growth. Long duration crop requires more water. Crop at seeding stage may require low amount of water. (Consumptive use is low). The water requirement will be more when the crop is at vegetative stage and it increases to a peak at fruiting stage

c) Rooting habit of crops. Crops with deeper root system will have greater the area of water absorption and hence more water is applied.

d) Drought resistant crops needs less water for example – pearl millet

2. Nature of soil

Storage capacity and availability of water in a soil is a function of physical properties like texture, structure etc.

i) Light sandy textured soil requires more water than clayey or retentive soil.

ii) The coarse soil needs more water than fine sand because of more infiltration

iii) Shallow soils require more frequent irrigation though in small amounts.

iv) Soil fertility also influences the water use by the plants – Crops in low fertile soils transpire more water per kg of plant tissue produced.

v) Drainage condition of the soil also influences the water requirement of crops. In water logged soils poor aeration and low inlet of water leads to less transpiration and uptake of water, leading to low crop yields.

3. Climatic factor

i) High temperature leads to more water requirement because of more evapotranspiration.

ii) Lower the humidity higher is the water need.

iii) Amount of sunshine influences the water requirement of crops. More sunshine cause more evaporation and transpiration and thus water utilization is more.

iv) The rainfall received influences the water requirement of crops.

v) Constant wind of moderate velocity ultimately increases evaporation which in turn increases the crop water need.

4. Type of irrigation

i) Method of irrigation – Flood irrigation requires more water.

ii) Method of conveyance of water from irrigation source – Seepage and evaporation losses depends on the nature of canals.

iii) Method of delivery/size of the stream – Slow delivery leads to more evaporation and percolation and thus more water requirement

5. Other factors

i) Quality of water

ii) Cultural practices adopted

iii) Method of land preparation

Critical of Periods of Water Requirement for Various Crops

Critical stages of water requirement: The stages of the crop at which the moisture stress considerably affect the yield of crops are known as *critical stages*

Table with details of periods of water requirement for various crops

S. No.	Crops	Critical Period (s)
1.	Rice	Tiller initiation, panicle primordial initation and flowering
2.	Wheat	Tillering, crown root initiation and boot and dough stage
3.	Maize	Tasseling, silking and milk ripe stage
4.	Sorghum	Knee heigh stage (25 – 30 cm), boot and milk ripe stage
5.	Cotton	Pre-flowering 3 weeks after flowering and boll formation
6.	Pearl millet	Heading and flowering
7.	Finger millet	Primordial initiation, flowering,
8.	Groundnut	Flowering, peg penetration and pod development
9.	Sesame	Blooming to maturity

10.	Sunflower	From rosette to flowering
11.	Soybean	Blooming to seed formation
12.	Castor	Entire growing period
13.	Sugarcane	Formative phase
14.	Tobacco	Immediately after transplanting, knee heigh (45 cm) upto full growth of leaves
15.	Grams	Flowering and late vegetative phase
16.	Beans	Flowering and pod setting
17.	Peas	Flowering and early pod formation
18.	Alfalfa	Immediately after cutting for hay, starting of flowering for seed production
19.	Onion	Bulb formation to maturity
20.	Tomato	Flowering and commencement of fruit set
21.	Chillies	Flowering
22.	Cabbage	Head formation until becoming firm
23.	Carrot	When root enlargement starts
24.	Potato	Tuber initiation to maturity
25.	Citrus	Flowering, fruit setting and enlargement stages
26.	Mango	Flowering, some water stress before flowering
27.	Coffee	Flowering, and fruit development some water stress before flowering generally held desirable
28.	Banana	Requires adequate moisture throughout the growth and fruit development ; particularly during periods of hot weather

Drainage

- Drainage is the artificial removal of water in excess of the quantity required for the crop.
- Drainage includes removal of excess water of both surface and subsurface in the root zone of crops.
- Irrigation and drainage go together and are not mutually exclusive.
- Irrigation aims at supplying optimum quantities of water throughout the crop period, whereas, drainage aims at removing excess quantity of water in a short time.
- Often, both may be required together to assure sustained and high level production of crops.

Role of drainage

- Draining the land provides conditions favorable for crop production.
- The greatest benefit of drainage relates to aeration.
- Good drainage facilitates the ready diffusion of oxygen to the root zone and escape of carbon dioxide from the root zone into the atmosphere.
- Several harmful gases also escape from the root zone into the atmosphere.
- The activity of aerobic organisms which influence the availability of nutrients such as nitrogen and sulphur to plants depends on soil aeration and hence, drainage improves aerobic organisms.
- Toxicity in acid soils due to excess iron and manganese is decreased by drainage (due to presence of oxygen in the root zone).
- Drainage permits roots to grow deeper and spread wider thereby increasing the volume of soil from which nutrients can be extracted.
- The removal of excess water helps in drying of the soil quickly and permits timeliness of field operations.
- The provision of a good drainage system permits the removal of excess salts in the soil or irrigation water and prevents their build up in the upper soil layers.

Questions

Fill the blanks

1. Artificial application of water to the soil for the purpose of crop production in supplement to rainfall and ground water contribution is called ________.
2. In __________irrigation Water is applied in the root zone of the crop.
3. Light sandy textured soil requires ________ water than clay or retentive soil
4. The stages of the crop at which the moisture stress considerably affect the yield of crops are known as ______________.
5. __________ is the artificial removal of water in excess of the quantity required for the crop.

Choose the correct answer

6. Ninety per cent of world's total irrigated area is under this irrigation method

a. Surface b. Subsurface

c. Drip d. Sprinkler

7. In this method of irrigation a set of furrows are completely skipped out from irrigation permanently.

a. Alternate furrow b. Skip furrow

c. All furrow d. Check basin

8. Water requirement – (Effective RF + Groundwater contribution) is

a. Irrigation requirement b. Water requirement

c. Consumptive use d. None

9. The irrigation method suitable for water scarcity and salt affected soils is

a. Check basin b. Border irigation

c. Drip d. Sprinkler irrigation

10. The irrigation in which fields are divided into several plots and water is guided from the stream through channels in a controlled manner.

a. Controlled flooding b. Uncontrolled flooding

c. Surface d. Sub surface

Answer the following

1. Irrigation scheduling
2. Alternate furrow irrigation
3. Advantages of drip irrigation
4. Water requirement of crop
5. Role of drainage.

14

Manures and Fertilizers

Growth is the development of a plant as a whole or of a specific organ. Besides the genetic factors, the environmental factors grouped as climatic factors and soil factors influence plant growth. The supply of mineral nutrient elements to the plants is discussed in this chapter.

A complete analysis of plants detects large number of elements. But only certain elements are essential. An element is said to be **essential** if the plant cannot complete its life cycle without it, and if the malady (deficiency) that develops in plants in its absence can be remedied only by that element.

Earlier 16 elements were considered as essential for plant growth. They are carbon, hydrogen, oxygen, nitrogen phosphorus, potassium, calcium, magnesium, sulphur, iron, manganese, zinc, copper, molybdenum, boron and chlorine. Recently, sodium, cobalt, vanadium, silica, selenium, gallium, aluminium and iodine are added to the above list. One or the other of these elements (recently added) has been found to be essential for a particular species of plants.

Carbon dioxide, water and molecular oxygen are the forms in which C., H and O are assimilated by plants. Others are taken up by plants from the soil. Nutrient uptake by plants accounts for about 10 percent of total dry weight of crops, the remaining percentage being water. The chemical symbol and the ionic forms in which the essential elements are absorbed by the plants are as follows.

Element	Symbol	Form (s) of absorption by plants
1. Carbon	C	CO_2
2. Hydrogen	H	H from H_2O
3. Oxygen	O	elemental O_2 and O_2 from H_2O
4. Nitrogen	N	NH_4^+, NO_3^- also as organic $CO(NH_2)_2$ and molecular nitrogen

5.	Phosphorus	P	HPO_4^{2-}, $H_2PO_4^-$ also as Nuclic acid, Phytin
6.	Potassium (Kalium)	K	K^+
7.	Caclcium	Ca	Ca^{++}
8.	Magnesium	Mg	Mg^{++}
9.	Sulphur	S	SO_3^{2-}, SO_4^{2-}
10.	Iron	Fe	Fe^{++}, Fe^{+++}
11.	Zinc	Zn	Zn^{++}
12.	Manganese	Mn	Mn^{++}
13.	Copper	Cu	Cu_2^{++}
14.	Boron	B	BO_3^{3-}
15.	Molybdenum	Mo	MoO_4^{2-}
16.	Chlorine	Cl	Cl^-
17.	Silicon	Si	$Si(OH)_4$
18.	Sodium	Na	Na^{2+}
19.	Cabalt	Co	Co^{2+}
20.	Vanadium	V	V^+

Classification of Essential Elements

Essential elements needed for the crop growth are broadly classified,

1. Based on the relative quantity that is normally present in plants

a. *Macro nutrients Major Nutrients / primary nutrients)* : C, H, O, N, P, K

b. *Secondary nutrients* : Ca, Mg, S

c. Micro Nutrients (Minor / Tertiary / Trace elements): Fe, Mn, Zn, Cu, Mo, B and C1. (Na, Se, Co, V, Ga, A1 and 1).

2. Based on their chemical nature

i) Metals : K, Ca, Mg, Fe, Zn, Mn, Cu, Co and V etc.,

ii) Non-Metals : C, H, O, N, P, S, B, Mo, Cl, Si, etc.,

iii) Cations : NH_4^+ K^+, Ca_2^+, Fe_2^+, Mg_2^+, Mn_2^+, Cu_2^+, Zn_2^+

iv) Anions : NO_3^-, HPO_4^{2-}, $H2P_{O4}^-$, SO_4^{2-}, BO_3^{3-}, MoO_4^{2-}, Cl^- etc.,

3. Based on general function

i) As a constituent of either organic or inorganic compunds – N, S, P, Ca, B, Fe and Mg.

ii) As an activator, cofactor in prosthetic group of enzyme systems – K, Mg, Ca, Fe, Zn, Mn, Cu, Mo, Na and C1.

iii) As a charge carrier in oxidation – reduction reactions – P, S, Fe, Mn, Cu, Mo.

iv) As an osmo (sis) regulator and for electron chemical equilibrium in cells – K, Na and Cl.

4. Based on the mobility in plants

i) Highly mobile : N, P, K

ii) Moderately mobile : Zn

iii) Less mobile : S, Fe, Cu, Mn, Cl, Mo.

iv) Immobile : Ca, B

Role of Nutrients in Crop Production

The role of nutrients, deficiency, control of deficiency and toxicity are given below :

Nutrient (Element)	Role of Nutrients	Deficiency symptoms of nutrients	Control of deficiency	Symptoms under excess nutrients
1. Nitrogen (N)	1. It is constituent of chlorophyll. 2. N makes plant dark green 3. It increases vegetative growth, protein content and cation exchange capacity in plant roots 4. Encourage the formation of good quality foliage	1. Lower leaves turn yellow 2. Growth of plant is stunted 3. Shedding of leaves and fruits	1. Use of nitrogen fertilizer in the soil 2. Foliar spray of urea	1. Blackening around tips of older leaves 2. Delays maturity 3. Encourages Lodging 4. Makes plant more susceptible to pests and diseases. 5. Poor root growth

2. Phosphorus (P)	1. Stimulates root growth and formation 2. Helps in cell division 3. Hasten maturity 4. Makes plant more tolerant to drought, cold, insects and diseases 5. Increase P and Ca in plants 6. Increase tillers and ratio of grain to straw in crop	1. Leaves become smaller in size 2. Leaves and stems become purple 3. Delay in maturity 4. Growth is stunted	1. Application of phosphatic fertilizers in the soil, e.g., super phosphate	1. Necrosiss and tip dieback 2. Interveinal chlorosis in younger leaves scorch of older leaves
3. Potasium (K)	1. K – helps in translocation 2. Imparts, vigour and growth to plants 3. Makes plant more tolerant to drought, cold insects and diseases. 4. Reduces lodging 5. Increases the availability of N and P 6. Increases the size of root and tuber	1. Margin of leaves turn brown and dry up 2. The older leaves develop brown colour 3. Stunted growth	1. Use of potassic fertilizer in the soil e.g., muriate of potash	Plants have luxury consumption hence not toxic
4. Calcium (Ca)	1. Promotes early root growth 2. Ca is constituent of cell	1. Terminal bud dies 2. Leaves become wrinkled 3. New leaves	1. use of calcium carbonate or calcium hydroxide in the soil	

	3. Increases stiffness of straw (stem) 4. Improves soil structure 5. Keeps soil neutral	shows symptoms	2. Use of gypsum	
5. Magnesium (Mg)	1. Constituent of chlorophyll 2. Increases photos ynthesis 3. Regulates uptake of nutrients 4. Promotes the formation of oils and fats	1. Vein of leaves remain green and inter veinal chlorosis 2. Symptoms on older leaves	1. Foliar application of magnesium sulphate (Epsum)	1. May induce K deficiency
6. Sulphur (S)	1. Helps in chlorophyll formation 2. Stimulates root growth seed formation and nodule formation 3. Encourages plant growth 4. S is constituent of enzymes and proteins	1. the whole leaf in plant has light green colour	1. Foliar application of sulphur or sulphate	1. Reduction in leaf size
7. Iron (Fe)	1. Helps in chlorophyll formation 2. Acts as an oxygen carrier 3. Helps in protein synthesis	1. Yellowing of new check leaves 2. Chlorosis	Spraying of 0.5% Ferrous sulphate on foliage	Bronzing of older leaves is common in low land rice grown under acid soils.
8. Manganese(Mn)	1. Acts as a catalyst in	1. Brown patch on leaves	1. Soil or foliar application of	1. Spots on the veins of the leaf

	oxidation reduction reaction 2. Act as activator of many enymes 3. Helps in chlorophyll synthesis	2. Reddening of leaves in cotton	manganese sulphate	blade and leaf sheath 2. Stunted plant
9. Boron (B)	1. Helps in uptake and utilization of calcium 2. Helps in protein synthesis	1. The leaves thicken and margin roll upward 2. younger leaves are dwarf 3. Top-rot diseases of tobacco	1. Foliar spray of boric acid or borax 2. Use of boron in soil	1. Inter veinal chlorosis at the tips of the older leaves along the margins 2. Leaves turn brown and dryup
10. copper (Cu)	1. Helps in oxidation –reduction reaction			
11. Molyb-denum (Mo)	1. Helps in focation of atmoshperic nitrogen by nodule bacteria in legume 2. Helps in protein synthesis	1. Petiole of the leaves remain intact but shedding of margin and other part of leaves 2. Curling of leaves	1. Soil or foliar application of sodium molybate or ammonium molybate	Not common
12. Chlorine (Cl)	1. Essential for photosyn-thesis process 2. Keeps osmotic pressure normal in cell sap	1. Yellowing of leaves (white plant)	1. Potassium chloride application in the soil	1. Burning of leaf tips or margins 2. Reduce leaf size
13. Zinc (Zn)	1. Constituent of a number of	1. White leaf become rusty-	1. Soil application of	1. Induces iron chlorosis

enzymes 2. Helps in formation of growth hormones 3. Act as catalyst in chlorophyll formation	brown in colour 2. Stunted growth	Zinc sulphate @ 25-50 Kg/ha. 2. Foliar application of 0.5% zinc sulphate

Nutrient Deficiency Symptoms of Crops

Nutrient deficiency symptoms of crops: Plant symptoms can be grouped into five types as follows

1. ***Chlorosis :*** Yellowing, either uniform or interveinal of plant tissue due to reduction in the chlorophyll formation process
2. ***Necrosis :*** Refers to death of plant tissue leading to dead spots
3. Lack of new growth or terminal growth resulting in ***" resetting"***
4. ***Accumulation of anthocyanin*** and an appearance of a reddish colour
5. Stunting or reduced growth with either normal or dark green colour or *yellowing,*

Nutrients are continuously removed from the soil by crops in addition to losses by leaching, volatilization and erosion. This nutrient is added to the soil by external sources to maintain soil fertility and sustainable production. Manure is the organic material derived form animal, human and plant residues which contains plant nutrients in complex organic forms. The major organic sources are manures are farm waste, cattle shed waste, human habitation waste, slaughter house waste, fishmeal, byproducts of agro-industries etc. The manures are bulky, concentrated, green and greenleaf manures depending on their volume and nutrient content. Of these two sources most widely used all over the world, one is organic in nature – the organic manures simply called manures and the other comprises the synthetic or naturally occurring chemical fertilizers simply called fertilizers.

Manures

Manures are plant and animal wastes that are used as source of plant nutrients. They release nutrients after their decomposition. Manures can be grouped into bulky organic manures and concentrated organic manures.

a) Bulky organic manures - Farm Yard Manure (FYM), compost from organic waste, night soil, sludge, sewage, green manures.

b. Concentrated organic manures - oilcakes (edible, non-edible), blood meal, fishmeal and bone meal.

Fertilizers

Fertilizers are industrially manufactured chemical containing plant nutrients. Nutrient content is higher in fertilizers than organic manures and nutrients are released almost immediately.

The fertilizers has three groups;

a) Straight fertilizers – supplies single nutrient Ex: Urea, Muriate of Potash

b) Complex fertilizers - supplies two or more nutrient Ex: 17:17:17 NPK complex

c) Mixed fertilizers- supplies two or more nutrient Ex: Groundnut mixture.

Role of Manures and Fertilizers

1. Organic manures bind the sandy soil and improve its water holding capacity.
2. Organic manures open the clayey soil and help in aeration for better root growth.
3. Organic manures add plant nutrients in small percentage and also add micronutrients, which are essential for plant growth.
4. Manures increase the microbial activity which helps in releasing plant nutrients to available form.
5. Organic manures should be incorporated before the sowing or planting because of slow release of nutrients.
6. Fertilizers play an important role in crop production as they supply large quantities of essential nutrient to crops
7. Fertilizers are manufactured in forms that are readily utilized by plants directly or after rapid transformation.
8. Fertilizers dose can be adjusted to suit the requirement as determined by soil testing.
9. Balanced application of nutrient based on crop requirement is possible by appropriate mixing of fertilizers.

10. Fertilizers applied as straight fertilizers (providing single nutrient) or complex and mixed fertilizers (supplies two or more nutrients) based on crop requirement.

Agronomic Interventions for Enhancing FUE

The following are the agronomic measures to improve the Fertilizer use efficiency (FUE).

1. Using best fertilizer source
2. Using adequate rate & diagnostic techniques
3. Usage of balanced fertilization
4. Integrated nutrient management
5. Utilization of residual nutrients

1. Using best fertilizer source

Identification of best source of fertilizer is pre-requisite for better crop production. Source of fertilizer depends on crop and variety, climatic and soil condition, availability of fertilizer, etc.

- Nitrogen: Ammoniacal or Nitrate
- Phosphorus: Water soluble or Citrate soluble
- Potassium: Muriate of potash
- Sulphur: Sulphate or Elemental S
- Multinutrient fertilizers: MAP, DAP, SSP, Nitrophosphates
- Multi-nutrient mixtures: Several combinations of NPK
- Fortified fertilizers: Neem-coated urea, Zincated urea, Boronated SSP, NPKS mix.

2. Using adequate rate & diagnostic techniques

The fertilizer recommendation must be in adequate quantity so as to meet the demand of crop at any point of growth. The fertilizer supply is made by diagnosing its requirement by any of the following method.

- State recommended generalized fertilizer dose or blanket recommendation
- Soil-test based fertilizer recommendations

- Soil-test crop response based recommendation
- Plant analysis for diagnosing nutrient deficiencies
- Chlorophyll meter and Leaf colour charts, etc.

3. Balanced fertilization

Balanced fertilization includes adequate supply of all essential nutrients, proper method of application, right time of application and nutrient interrelationships.

a) Adequate supply of all essential nutrients

Due to more concentration and application on primary nutrients (NPK), soils developed deficiency symptoms for secondary and micro-nutrients. Hence, ignored elements must be added with the NPK (may be in minor quantity) to get higher yields in crops. Experimental results shown that about 20-25 kg of micro-nutrient application or two foliar sprays increases the yield of crops up to 20%.

b) Proper method

N and K can be applied as broadcasting and band placement. Water soluble P fertilizers are preferred to apply as band placement in neutral & alkaline soils. Citrate soluble P fertilizers are applied as broadcast method in acidic soils. Sulphate forms of S fertilizers are applied as broadcasting or band placement, whereas, elemental S and pyrite are applied as broadcasting method. Micronutrients are applied in minor quantity as foliar sprays and water soluble fertilizers are applied in fertigation.

c) Right time: (according to physiology of crop)

- Upland crops - 2 splits (seeding, 3-5 weeks after first dose)
- Flooded rice - 3 splits (Transplanting, 3 and 6 weeks after first dose)

d) Nutrient interrelationships

Antagonistic nature of fertilizers is to be considered while applying into the soil. Some of the fertilizer application in excess, cause loss of yield and quality of crops. Ex. Application of excessive 120 kg P ha^{-1} created an imbalance and reduced the seed and oil yields in soybean compared to 80 kg P ha^{-1}.

4. Integrated nutrient management

Organic manures, crop residues, green manures, bio-fertilizers etc. are to be blended in right manner along with inorganic fertilizers to meet the crop demand. All the possible and available organic sources are to be utilized efficiently to reduce the usage of inorganic fertilizers.

5. Utilization of residual nutrients

Some of the strategies to utilize the crop residues in efficient manner are :

- Knowledge on climatic conditions & carry-over effects of residues.
- Blending rightly on cereal-legume rotations
- Mixing shallow-deep rooted crop rotations

Classification of Manures

A. Bulky organic manures

i) Farm Yard Manure:

a) Cattle manures

b) Sheep manures

c) Poultry manures

ii) Compost:

a) Village / rural compost is made from farm-wastes

b) Town / urban compost is made from town refuses

iii) *Sewage and sludge*

B. Concentrated organic manures

1. Oil cakes

a) Edible oil cakes (i.e., used for cattle feeding)

i) Mustard cake ii) Groundnut cake

iii) Sesame cake iv) Linseed cake

b) Non edible oil cakes (i.e. used as manures)

i) Castor cake ii) Neem cake

iii) Sunflower cake iv) Mahua cake

v) Karanja cake

2. Slaughter House wastes

i) Blood meal and ii) Bone meal

3. Fish meal

4. Guano

Material obtained from the excreta and dead bodies of sea brid

C. Green manures

a) *Leguminous plant* (example Sunnhemp, Sesbania sp., Mungbean, Cowpea, Guar, Senji, Berseem)

b) *Non-leguminous plant* (example Sorghum, Pearl millet, Maize, Sunflower)

D. Green leaf manures

Green leaves of trees like neem, pungam, glyricidia, vadhanarayana etc.

Organic manures include plant and animal by-products such as oil cakes fish manures and dried blood from slaughter houses. Before their organic nitrogen used by the crops it is converted through bacterial action into readily usable ammoniacal – N and nitrate – N. These manures are therefore, relatively slow acting, but they supply available N for a longer period.

Bulky organic manures are those manures which are generally bulk in quantities and poor in plant nutrients (low quantities of plant nutrients) Example – Farm yard manure, Compost, Sewage and sludge etc.,

i) Farm Yard Manure (FYM): It is the manure produced in the farm which is made up of excreta (dung and urine) of farm animals, the bedding materials provided for them and miscellaneous farm and house hold wastes. Straw, peat and saw dust, dry leaves etc., are used as bedding material for farm animal and accounts to 3 – 4 kg per animal per day. The bedding material is called '*litter*' and it absorbs urine voided by animals. It is not a standardized product and its value depends on the kind of feed fed to the animal, the amount of straw used and the manner of storage. In general FYM contains 0.8% N, 0.41% P_2O-$_5$ and 0.74% K_2O.

The excreta of horses and sheep are drier than other and do not get compacted in the heap. There is considerable aeration, bacterial activity and rise in temperature in the manure. They are therefore called '*hot manures*' in the temperate countries.

Pig and cattle manure contain more moisture and compacted in the manure pit. Their decomposition is not so vigorous as that of hot manures and the rise of temperature is also low. Therefore pig and cattle manures are called "*cold manures*". The decomposition of cattle manures may be slower comparatively under temperate regions but it is rapid enough under tropical condition.

ii) Compost: It is a manure derived from decomposed plant residues usually made by fermenting waste plant materials heaped or put in a pit usually in alternate layers with a view to bring the plant nutrients in a more readily available form.

Super Compost: Compost fortified with superphosphate is called as a super compost.

Starters are the materials added to the composting organic wastes which provide the decomposing organism. Pig dung slurry is a valuable starter and provides necessary organisms. Even cowdung slurry can be used as starter.

Generally ammonium sulphate and super phosphate are added to the layers at the time of furrowing the composting heap to enrich nitrogen and phosphorus status of compost respectively. Fertilizers accelerate and hasten the decomposition of organic matter or wastes.

iii) Sewage and Sludge: In cities human excreta are flushed out with large quantities of water which is known as sludge. It contains two components, one is solid portion called sludge and another is liquid portion called sewage water. In general the sludges are rich in N and P and low in K. The sewage water is used for irrigation after proper treatments.

Nutrient Content of the Bulky Organic Manures

Manure	Percentage composition of		
	N	P_2O_5	K_2O
Cattle dung	0.40	0.20	0.20
Cattle urine	1.00	-	1.35
Sheep and goat dung	0.75	0.50	0.45
Sheep and goat urine	1.35	0.05	2.10
Sheep and goat manure	3.00	1.00	2.00

Poultry manure	3.03	2.63	1.40
Horse manure	2.00	1.50	1.50
Horse urine	1.35	-	1.25
Pig dung	0.60	0.50	0.40
Pig urine	1.10	0.10	0.45
Farm litter compost	0.50	0.15	0.50
Rural compost	1.22	1.08	1.47
Town compost	1.40	1.00	1.40
Waterhyacinth compost	2.00	1.00	2.30
Vermicompost	3.00	1.00	1.50
Night soil	5.50	4.00	2.00
Paddy straw	1.50	1.34	3.37
Sugarcane trash	2.73	1.81	1.31
Sewage sludge	1.5-3.5	0.75-4.00	0.3-0.6

Concentrated organic manures are those manures which are rich in particular nutrients (N) but relatively having low volume of organic materials. Example – Oil cakes, blood, bone and fish meal, press mud etc.,

Oil cakes: Oil cake is the residue left after the oil is extracted from oil containing seed. The manurial values of oil cake lie mainly in its nitrogen contribution though it is in small quantities. The nitrogen content varies between 3 and 9%. The C:N ratio is 3 to 15 for most of the oil cakes.

Nutrient content of some concentrated organic manures

Manure	Percentage composition of		
	N	P_2O_5	K_2O
Castor cake	4.0-4.4	1.9	1.4
Groundnut cake	6.5-7.5	1.3	1.5
Cotton seed cake (decorticated)	6.9	3.1	1.6
Cotton seed cake (undecorticated)	3.6	2.5	1.6
Linseed cake	5.6	1.4	1.3
Coconut cake	3.4	1.9	1.9
Neem cake	5.2-5.6	1.1	1.5
Safflower cake (decordicated)	7.9	2.2	1.9
Safflower (undecorticated)	4.9	1.4	1.2

Sesamum cake	4.7-6.2	2.1	1.3
Mahua cake	2.5	0.8	1.9
Niger cake	4.7	1.8	1.3
Pungam cake	4.0	1.0	1.3
Raw bone meal	4.0	20-25	-
Steamed bone meal	4.7	25-30	-
Basic slag	4.0	1.0	1.3
Fish meal	4-10	3-9	1.5
Blood meal	10-12	1-2	1.0
Meat meal	9-11	3.5	-
Horn and hoof meal	10-15	1	-
Press mud	1-1.5	4-5	2-7
Guano (Peruvian bird)	11-16	8-12	2-3

Green manure and green leaf manure

Green manuring is the act of growing of quick growing crop preferably legumes and ploughing *in situ* and incorporated into the soil. Whereas green leaf manuring is incorporation of green matter into the soil transported from elsewhere. The percentage N of some of the green/ green leaf manures in give below.

Nutrient content of green manure crops and green leaf manures

Plant	Scientific name	Nutrient content (%) on air		
		N	P_2O_5	K_2O
Green manure				
Sunnhemp	*Crotolaria juncea*	2.30	0.50	1.30
Manila agathi	*Sesbania rostrata*	3.30	0.60	1.20
Daincha	*Sesbania aculeata*	3.20	0.60	1.20
Pillipesara	*Phaseolus trilobus*	2.80	0.50	1.15
Sesbania	*Sesbania speciosa*	2.71	0.53	2.21
Kolinji	*Tephrosia purpurea*	3.10	0.52	1.18
Green Leaf manure				
Glyricidia	*Glyricidia sepium*	2.76	0.28	4.60
Pongamia	*Pongamia glabra*	3.31	0.44	2.39
Neem	*Azadiracta indica*	2.83	0.28	0.35

Gulmohar	*Delonix regia*	2.76	0.46	0.50
Vadanarayanan	*Delonix elata*	3.51	0.31	0.43
Subabul	*Leucaena leucocephala*	3.50	0.48	0.81
Peltophorum	*Peltophorum ferrugenium*	2.63	0.37	0.50
Weeds				
Parthenium	*Parthenium hystorophorus*	2.68	0.68	1.45
Water hyacinth	*Eichhornia crassipes*	3.01	0.90	0.15
Sarannai	*Trianthema portulacastrum*	2.64	0.43	1.30
Aduthoda	*Aduthoda vesica*	1.32	0.38	0.15
Ipomea	*Ipomoea cornea*	2.01	0.33	0.40
Calotrophis	*Calotrophis gigantean*	2.06	0.54	0.31
Cassia	*Cassia fistula*	1.60	0.24	1.20

Stem nodulating green manure

Leguminous green manure plants produce root nodules and fix atmospheric N. *Sesbania rostrata* produces nodules on their stem besides root nodulation. This special feature adds their green manurial value. It is tropical legume of Senegal origin and thrives well under flooded and water logged conditions. It is capable of producing 22 tonnes of fresh biomass and could accumulate 150 kg N / ha in 45 days. It contains 3.3% N.

Benefits of organic manures application to soil

Organic manures, i) supply plant nutrients including micro nutrients ii) improves physical properties of the soil, water holding capacity, hydraulic conductivity, infiltration capacity of the soil, iii) CO_2 released during decomposition combines with water and forms carbonic acid and act as CO_2 fertilizer, iv) supplies energy (food) for microbes v) increase availability of nutrients and improve soil fertility vi) green manures have the additional advantage of fixing atmospheric nitrogen leading to nitrogen economy in crop production and vii) green manures with draw nutrients from lower layers and concentrated them in the surface soil for the use of succeeding crop.

Importance of daincha in reclamation of saline and alkali soils

Green manuring practice in sodic soil has an unique importance since it adds acids in the reclamation process, besides improving the fertility status of the soil. Usually the fertility status of sodic soil is very poor because of its high pH and exchangeable sodium percentage. The soil organic matter content, a measure of available nitrogen, is very low i.e. 0.1 to 0.5% in sodic soil because sodium carbonate and sodium bicarbonate salts in solution dissolve the humus. Further the available nitrogen is much lower in the subsoil layers of the sodic soils.

Reclamation of alkali soils basically involves replacing Na on the exchange complex with more favourable cations. The solubility of lime, which is always present in alkali soils in significant amounts, is very low, because the potassium content of alkali soil is high. There is an intimate relationship between soil pH, partial pressure of CO_2 and calcium ion activity in calcareous alkali soils. Increase in CO_2 production in the soil enables to increase the soluble Ca status of soils. This in turn, replace exchangeable Na, resulting in the improvement of alkali soils.

Soil incorporation of easily decomposable plant material results in increased and rapid production of CO_2. For this reason, green manuring has been suggested as an important management practice for the reclamation of alkali soils.

Sesbania aculeata and *Delonix elata* are very effective green manures and green leaf manures respectively used for reclamation of sodic soils.

Daincha (*Sesbania aculeata*) is highly resistant to both drought and water stagnation and salinity and alkalinity. It can be grown in soils with pH 4.5 to 9.5. It produces green matter of 20 t/ha in 90 days. Daincha contains 3.2% N and 34% Ca on dry weight basis which helps to replace Na from sodic soils.

The acid juice (pH 4.0) and high seed protein content (58%) seems to be the cause of its resistance to sodicity stress.

During the reclamation of sodic soils gypsum @ 50% of gypsum requirement (GR) has to be spread uniformly over the field. The surface soil is to be ploughed to mix the gypsum in the sodic soil. Irrigate the field with 10-15 cm depth of water and maintain the same water depth for 3-4 days. At this stage, the sodium content in clay particles are replaced by the calcium ions from the gypsum, allowing the sodium to wash out of the field as Leachate. The field has to be kept with stagnant water for 3 to 4 times after each drainage process. Apply the vadanarayanan (*Delonix elata*) leaves and daincha @ 5 t/ha without allowing the soil to dry. After four to five days of incorporation of green leaves, the field crop like rice preferably a tolerant variety CO 43, TRY1 etc.,

Fertilizers are synthetic (commercially manufactured) or naturally occurring chemical compounds either dry solid or liquid that added to the soil to supply one or more plant nutrients for crop growth.

Classification of Fertilizers

The fertilizers are classified based on whether the fertilizer supplies a single or more than one nutrient, their chemical nature and commercial mode of supply as straight, compound, complex and mixed.

1. **Straight fertilizers:** When a fertilizer contains and is used for supplying a single nutrient, it is called a straight fertilizer. This is further classified as nitrogenous, phosphatic and potassic fertilizers depending on the specific macro nutrient present in the fertilizer.

A) **Nitrogenous fertilizers:** Fertilizers containing N as major nutrient. It may be either a nitrate or ammonium or amide fertilizer depending on the form of nitrogen present.

The nutrient composition of different N fertilizers are listed below

Sources form	Nutrient content (Percent) Available							
	N	P_2O_5	K_2O	CaO	MgO	S	Cl	
Ammonium sulphate	20.6	-	-	-	-	24	-	NH_4^+
Ammonium chloride	25-26	-	-	-	-	-	66	"
Ammonium nitrate	33-34	-	-	-	-	-	-	
Amm. sulphate nitrate	26.0	-	-	-	-	-	-	"
Anhydrous ammonia	82.0	-	-	-	-	-	-	"
Calcium amm. nitrate (CAN)	35	-	-	8.1	4.5	-	-	"
Calcium nitrate	15	-	-	34	-	-	-	NO_3
Sodium nitrate	16	-	-	-	-	-	-	"
Urea	46	-	-	-	-	-	-	Amide
Calcium cynamide	21	-	-	-	-	-	-	"

B) Phosphatic fertilizers: They are classified into three groups, based on the solubility of phosphate contained in the fertilizer.

i) Water soluble phosphate (Mono calcium phosphate) $Ca(H_2PO_4)_2$

a) Single super phosphate 16% $Ca(H_2PO_4)_2' H_2O$

b) Double super phosphate 32% $2Ca (H_2PO_4)_2' H_2O$

c) Trible super phosphate 48% $3Ca (H_2PO_4)_2' H_2O$

ii) Citric acid soluble phosphate (Di-calcium phosphate) $Ca(H_2PO_4)_2$

a) Basic slag (CaO)3 $P_2O_5SiO_2$ 14-18% (By product from steel industry)

b) Di-Calcium Phosphate $Ca_2 (H_2PO_4)_2$ 34-39%

iii) Insloube phosphate (Tri-calcium phosphate)$Ca_3(PO_4)_2$

a) Rock phosphate 20-40% $Ca_3(PO_4)_2 CaF_2$

b) Raw bone meal 20-25% Ca $(PO_4)_3 CaF_2$ (2 - 4%N)

c) Steamed bone meal 22% - 30%

C) Potassic fertilizers

Muriate of Potash (KCI)	60%
Sulphate of potash (K_2SO_4)	48 - 52%
Potassium nitrate (KNO_3)	48% (nitrogen 13%)
Schoenite ($K_2SO_4{'}$ $MgSO_4$) $6H_2O$	22 - 24%

2. **Compound fertilizers**: are the commercial fertilizers in which two or more primary nutrients or chemically combined. For example - DAP. DAP contains 18% N and 46% P_2O_{-5}

Fertilizer	N	P_2O_5	K_2O
Di anmmonium phosphate (DAP)	18	46	-
Mono ammonium phosphate	11	48	-
Urea ammonium phosphate	28	28	-
Ammonium phosphate	16	20	-

2. **Complex fertilizers:** are the commercial fertilizers containing 2atleast two or more of the primary essential nutrients at higher concentration in one compound. The nutrients in complex fertilizers are physically mixed.

Fertilizer	N	P_2O_5	K_2O
Complex fertilizers	17	17	17 (MF)
	14	28	14 (MF)
	10	26	26 (IFFCO)
	12	32	16 (IFFCO)
	14	36	12 (IFFCO)
Nitro-phosphate-potash	15	15	15
Gromor	14	35	14

- MF- Madras Fertiligers Ltd.
- IFFCO - Indian Farmers Fertilizers Co-operative Ltd.

3) Mixed fertilizers / Fertilizers mixtures

They are physical mixtures of two or more straight fertilizers. Sometimes a complex fertilizer is also used as one of the ingredients. The mixing is done mechanically. The fertilizer mixtures are usually in powder form but techniques have been developed for granulation of mixtures so that each grain will contain all the nutrients mixed in the mixture.

Standard fertilizer mixtures for specific crops

Mixture No.	Composition %			Crops
	N	P_2O_5	K_2O	
1.	14	7	0	Millets, rainfed cotton
2.	12	6	6	Fruit crops
3.	6	6	12	Fruit crops
4.	6	12	6	Potato, paddy
5.	9	9	9	Paddy, millets
6.	15	0	15	Top dressing for paddy, millets
7.	4	8	12	Groundnut
8.	6	6	18	Banana
9.	10	0	30	Banana
10.	15	5	5	Sugarcane
11.	16	4	4	Sugarcane
12.	16	0	12	Top dressing mixture
13.	10	10	0	Sorghum, pearl millet
14.	14	4	12	Coconut
15.	15	25	15	Paddy, millets, vegetables
16.	20	0	10	Sugarcane
17.	30	30	50	Tea
18.	17	17	17	Paddy

Salts Containing Secondary Nutrients:Calcium, sulphur and magnesium are termed as secondary nutrients since they are required comparatively less in quantity than primay nutrients (N,P,K) but more than micro nutrients. They are added to the soil through some fertilizers, like ammonium sulphate, calcium ammonium nitrate and phosphatic fertilizers. Commercial fertilizers containing these secondary nutrients are as follow :

i) Magnesium sulphate (Epsum) – 9.6% Mg and 13% S

ii) Calcium sulphate (Gypsum) – 19% Ca and 23% SO_4

Salts Containing Micro Nutrients: Copper, zinc, boron, manganese, iron molybdenum, chlorine are termed as micro nutrients since they are required in micro quantities. They are added to the soil through some commercial fertilizers as follows :

S.No.	Salt	Formula	Nutrient content
1.	Copper sulphate	$CuSO_4 5H_2O$	25 to 35% Cu
2.	Zinc sulphate	$ZnSO_4 7H_2O$	22 to 35% Zn
3.	Borax (sodium borate)	$Na_2B_4O_7 10H_2O$	10.6% B
4.	Manganese sulphate	$MnSO_4 4H_2O$	23% Mn
5.	Sodium molybdate	$Na_2MoO_4 2H_2O$	37 to 39% Mo
6.	Ammonium molybdate	$(NH_4)_6 Mo O_4 4H_2O$	54% Mo
7.	Ferrous sulphate	$FeSO_4 7H_2O$	20% Fe

Bio fertilizers

Bio fertilizers are the living organisms capable of fixing atmospheric nitrogen or making native soil nutrients available to crops. Atmospheric nitrogen is fixed effectively by the microorganisms either in symbiotic association with plant system (example Rhizobium, Azolla) or in associative symbiosis (Azospirillum) or in free living system (Azotobacter, phosphobacterium, blue green algae) or in mycrorrhizal symbiosis (VAM fungi)

a) Rhizobium

Rhizobium bacteria can fix atmospheric nitrogen symbiotically. They live in the nodules of host plants belonging to the family leguminoceae. The quantities of nitrogen fixed by Rhizobia differ with the rhizobial strain, the host plant and the environmental conditions under which the two develop. The species of the genus Rhizobium are numerous and require certain host plants. Fox example the bacteria that live symbiotically with soybean will not do so with alfalfa. A list of common legumes and the rhizobial strains by which they are inoculated is given in the table below.

Classification of Rhizobium – Legume associations

	Rhizobium species	Legumes
1.	*Rhizobium meliloti*	Alfalfa (Lucerne)
2.	*R. trifolii*	Clover
3.	*R. leguminosarum*	Peas
4.	*R. phaseoli*	Beans
5.	*R. lupine*	Lupine
6.	*R. japonicum*	Soybean
7.	*Rhizobium sp.*	Cowpea

Fixation of nitrogen by the leguminous plants will be at maximum only when the level of available soil nitrogen is at the minimum. It is sometimes advisable to include a small amount of nitrogen in the fertilizer of legume crops at sowing time (as a starter dose) to ensure that the young seeding will have an adequate supply until the rhizobia can become established. Larger quantity of nitrogen or continued applications of nitrogen, however reduce the activity of the rhizobia and therefore they are generally uneconomical. Rhizobial inoculation was found to fix 15 to 35 kg N per ha in a season on different pulse crops. Rhizobial inoculation can save up to 25% N fertilizer application to crops.

b) Azolla

It is a small water fern of world wide distribution under natural conditions. It contains the heterocystous bluegreen algae *Anabaena azollae* as a symbiont in an enclosed chamber in the dorsal leaf lobes. Azolla derives all its total nitrogen requirement by the symbiotic association with the algae. The *Azolla-Anabaena system* is agronomically most signification plant algal association and this is being employed as a nitrogen source for rice culture.

There are six species of Azolla. They are *Azolla caroliniana, Azolla filiculoides, Azolla mexicana, Azolla nilotica, Azolla microphylla* and *Azolla pinnata.*

Azolla contains 3.1-4.2 % N; 0.16% P_2O_5 and 0.18 % K_2O on dry weight basis.

c) Azospirillum

This bacterium is associated with cereals like rice, sorghum, maize, cumbu, ragi, thenai and other minor millets and also for cotton, sugarcane, oilseeds and fodder grasses. These bacteria colonizing in the roots not only remain on the root surface, but also a sizeable proportion of them penetrates into the root tissues and lives in harmony with the plants. They do not, however, produce any visible nodules or out growth on the root tissue. In the absence of any plant, azospirillum live in the soil just like any other micro organism saprophytically, however, when a suitable crop is raised, they are attracted towards the root system, where they colonize and grow in almost a symbiotic manner.

d) Azotobacter

The beneficial effects of azotobacter on plants was associated (non-symbiotically) not only with the process of nitrogen fixation but also with the synthesis of complex of biologically active compounds such as nicotinic acid, pyridoxin, biotin, gibberellin and probably other compounds which stimulate the germination of seeds and accelerate plant growth. Azotobacter population in soil or near the root zone of crops (Rhizosphere) is very low when compared to other soil bacteria. The nitrogen fixation potential of this bacterium is also not very high and appreciable (20 to 30 kg of N per ha per year). A fairly high population is required for substantial nitrogen fixation. Enoromous energy is required by azotobacter for nitrogen fixation. The possible source of energy for azotobacter is the soil organic matter. The energy generated during the utilization of organic matter is used for nitrogen fixation.

e) Blue green algae

The blue green algae occur under a wide range of environmental conditions. They are completely autotropic and require light, water, free nitrogen (N_2), carbon di oxide (CO_2) and salts containing the essential mineral elements. They play a major role in the nitrogen economy of paddy soils in tropical countries. Different algal species available are,

1. Tolypothric tenuis
2. Aulosira
3. Nostoc
4. Anabaena
5. Plectonema
6. Chorococcum
7. Chlorococcus

f) Phosphobacterium

In most of the acid and clayey soils, the applied phosphorus either as super phosphate or mussoriphos will not be available to crops due to fixation. It is essential to use the phosphobacteria (free living bacteria for proper solublisation of fixed P and release them in the available form for the crop to takeup for its growth. Dual inoculation of the phosphobacteria with rhizobium or azospirillum can provide both N and P to the crop.

g) Mycorrhiza (VAM)

Vesicular Arbuscular Mycorrhiza (VAM) is a fungi used as bio-fertilizer. The mycorrhizal symbiosis is an intimate association between plant root system and certain group of soil fungi. The plant provides carbon as energy source to the fungus which in turn help the plant in better uptake of nutrients (especially P).

The VAM fungi form either a mantle of hyphae around the root or penetrate inside the roots spreading intra or intercellularly in the cortical region. The fungal mycelium also extends several centimeter, away from the root in the soil. The area that the plant can explore for nutrients thus greatly increase due to colonization of plant roots by the mycorrhizal fungi. The development of mycorrhizal network is much more in soils with low fertility. In nutrient rich soils, there is very little extension of mycelial network. The mycelial growth is confined to the close proximity of roots. Mycorrhiza increases crop yield, protect against certain root pathogen, helps in uptake of P, Cu, Zn and B and increases tolerance to environmental stress.

Factors Affecting Manures and Fertilizers Use

Major factors influencing the selection, quantity, time and method of application of manures and fertilizers are,

1. Soil factors

They most important factors are, soil physical condition (texture), soil fertility and soil reaction.

i) Poor physical condition of the soil leads to poor plant growth due to impeded drainage, restricted aeration and unfavourable soil temperature. In this condition nutrients will not be used efficiency.

ii) Optimum soil moisture regime is essential for efficient use of fertilizers by crops.

iii) The availability of nutrients is poor in coarse textured soil when compared to fine textured soils. The coarse textured soil needs more frequent application of fertilizers when compared to heavy textured soil.

iv) The higher the fertility of soil, the lower is the response to manures and fertilizers.

v) When the organic matter of the soil is higher, the response to fertilizer by crops is more.

vi) Soil reaction is important for selection of right type of fertilizers Rock Phosphate advantageous in acid soils to avoid fixation of P.

2. Crop factors

i) The response of crop to fertilizers varies with the nature of crop and variety of the crop.

ii) The fertilizer responsiveness of a plant depends on the cation exchange capacity (CEC) of the roots. The root CEC of dicotyledonous plants is much higher than that of monocotyledonous plants. Plants with higher CEC absorb more of divalent cations (Ca, Mg) whereas plants with low CEC absorb more of monovalent ions (K, Na).

iii) The ability of the crop to absorb nutrients from the soil depends upon the size of the root system (root length and spread) and characteristics like root surface and root hair density etc. Large ramifying root system absorb more nutrients.

iv) The association of mycorrhizal fungi with the roots of plants grown under conditions of low soil fertility, increases the ability of plants to absorb nutrients such as P, K, Cu and Zn. Normally N, P and complete fertilizer application reduce the presence and activity of mycorrhiza.

3. Agronomic factors

Fertilizer responsiveness of crops depends on timely sowing, proper spacing, proper dose, time and method of fertilizer application.

4. Other factors

i) *Climatic factors*: Under drought and excess moisture condition, foliar spray can be recommended. In high rainfall area, split application of fertilizers and application of slow release nitrogenous fertilizers are recommended.

ii) *Yield goal*: The economic yield or potential yield or targeted yield decides the quantity of manures and fertilizers application. For higher crop yield optimum or maximum amount of fertilizers are to be applied.

iii) *Cost of fertilizers*: Not only the cost of fertilizers and manures but also the cost of together produce decide the quantity of manures and fertilizer to be applied i.e., depend on the profit from the crop. It may be maximum profit or maximum rate of return per rupee invested.

iv) *Availability of manures and fertilizers*: Timely availability of manures and fertilizers, transport facility and labour for application decides the quantity. Now a days, manures are not available to the required level due to various reasons.

Slow release fertilizers are developed to prevent the loss of nutrients by leaching and nitrification. It releases nutrients slowly and uniformly and increases the fertilizers use efficiency. Examples – Neem coated Urea Sulphur coated Urea, Lac coated Urea, Tar coated Urea, N-Serve, Isobutylidine di Urea (IBDU), Thio Urea etc.

Time of Application of Manures and Fertilizers

1. ***Before preparatory tillage***: Bulky organic manures, green manures, soil amendments and soil conditions are applied before preparatory tillage for thorough mixing with the soil.
2. ***Basal dressing***: Application of manures and fertilizers before last ploughign/puddlings or before sowing or planting.
3. ***At sowing or planting***: Concentrated organic manures, readily soluble and higher mobile fertilizers, slow release fertilizers, starter dose of N fertilizer to legume crops and fertilizer for specific nutrient deficient soil are applied during this time.
4. ***Top dressing:*** It is the application of manures and fertilizers to the established crop within crop duration. Top dressing may be done to the soil or to the foliage. Split application of nitrogen and potassium is done throughout the cropping period to increase the fertilizer use efficiency.

Method of Application of Manures and Fertilizers

The choice of method and time of fertilizer application depends on the form and amount of fertilizer, convenience of the farmer, the efficiency and safety of fertilizer application.

II Solid form

1. ***Broadcasting:*** The manures and fertilizers and fertilizers are scattered uniformly over the field before planting the crop and are incorporated by tilling or cultivating.

2. ***Drilling and Placement*:** Fertilizers are placed in the soil furrows formed at the desired depth. Placement can be done by the following ways.

   i) *Plough sole placement*: In this method of fertilizers are applied or dropped in the plough sole, which will be covered by the plough during the opening of adjacent furrow.

   ii) *Deep placement*: Fertilizers or manures are placed at the bottom of the top soil at a depth of 10-12 cm, especially in the puddle rice soil.

   iii) *Sub soil application*: Fertilizers are applied in the subsoil especially for tree crops and orchard crops at a depth above 15 cm.

3. ***Location or spot application*:** Fertilizer are placed in the root zone or the spot near the roots from which roots can absorb easily.

   i) *Contact of Drill placement*: Fertilizers or manures are placed at the time of drilling for placing the seeds. Fertilizers are manures will have good contact with the seeds or seedlings.

   ii) *Band placement*: This is the placement of manures or fertilizers or both in bands on the side or both sides of the row at about 5 cm away from the seed or plant in any direction. Such band placement is of three types.

      a) *Hill placement*: In widely spaced crops, like cotton, castor, cucurbits fertilizers or manures are applied on both sides of plants only but not continuously along the row.

      b) *Row placement*: In widely spaced crops between rows (Example – Sugarcane, maize, tobacco, potato) manures or fertilizers are placed on one or both sides of the row in continuous bands.

      c) *Circular placement*: Application of manures and fertilizers around the hill or the trunk of fruit tree crops in the active root zone.

   iii) *Pocket placement:* of fertilizers deep in soil to increase its efficiency. Especially for the sugarcane pocket placement is done. Fertilizers are put in 2 to 3 pockets opened around every hill by means of a sharp stick.

iv) *Side dressing*: refers to hill and ring placement of manures or fertilizers. It consists of spreading the fertilizer between the rows or around the plants

v) *Pellet application:* Nitrogen fertilizers are pelleted like mud ball or urea super granules (USG) and placed deep (10 cm) into the saturated soils (reduced zone) of wet land rice to avoid nitrogen loss from applied fertilizers.

Generally placement of fertilizer is done for three reasons.

i) Efficient use of plant nutrients from plant emergence to maturity.

ii) To avoid the fixation of phosphate in acid soils.

iii) Convenience to the grower.

II Liquid form

1. *Foliar application* : It refers to spraying of fertilizer solution on the foliage of plants for quick recovery from the deficiency (either N or S).

2. *Fertigation*: It is the application of fertilizer dissolved in irrigation water in either open or closed system i.e., lined or unlined open ditches and sprinkler or trickle systems respectively.

3. *Starter solutions*: They are solutions of fertilizers prepared in low concentrations which are used for soaking seeds, dipping roots, spraying on seedlings etc. nutrient deficient areas for early establishment and growth.

4. *Direct application to the soil*: Liquid fertilizers like anhydrous ammonia are applied directly to the soil with special injecting equipments. Liquid manures such as urine, sewage water and cattle shed washing are directly let in to the field.

Integrated Nutrient Management (INM)

Judicious combination of inorganic, organic and bio-fertilizers which replenishes the soil nutrients removed by the crops is referred as integrated nutrient management system.

Concept of INM is to integrate the nutrient sources and methods of organic and inorganic nutrient application to maintain soil fertility and productivity i.e., the complementary use of chemical fertilizers, organic manures and bio-fertilizers to solve the problems of nutrient supply, soil productivity and environment.

Developing an INM system for a particular crop sequence to a specific location requires a thorough understanding of (i) the effects of previous crop (ii) contribution of legume in the cropping system (iii) residual effect of fertilizers and (iv) direct, residual and cumulative effect of organic manures for supplementing and complementing the use of chemical fertilizers.

The main components of the N supply system are the organic manures green manures, crop residues, crop rotation and inter cropping involving legumes and cereals, bio-fertilizers including rhizobium, azotobacter, azospirillum, phosphorus solubilizing micro organisms like mycorrhizal fungi, azolla, blue green algae and cyanobacteria. All these can serve as an important supplementary source of nutrients along with the chemical fertilizers.

Thus, INM is environmentally non-degradable, technically appropriate economically viable and socially acceptable.

Questions

Fill the blanks

1. An element is said to be ___________ if the plant cannot complete its life cycle without it.
2. The fertilizer having the highest N_2 content is __________
3. Yellowing, either uniform or interveinal of plant tissue due to reduction in the chlorophyll formation process is termed as__________________
4. Adequate supply of all essential nutrients at the right time is called as ____________ fertilization
5. _______________ is a very effective green leaf manures used for reclamation of sodic soils.

Choose the correct answer

6. Potassium content in Muriate of Potash is

   a. 46.0 % b. 18.0 %

   c. 60.0 % d. 24.0 %

7. *Sunnhemp* is a

   a. Green manure b. Green leaf manure

   c. Fibre crop d. Pulse crop

8. Phosphobacteria is a
   a. Biocontrol agent
   b. Bioherbicide
   c. Biofertilizer
   d. Manure
9. Application of fertilizers with irrigation water is known as
   a. Sprinkler irrigation
   b. Fertigation
   c. Cabligation
   d. Drip irrigation
10. The fertilizer that contains high amount of nitrogen is
   a. Ammonium nitrate
   b. Ammonium sulphate
   c. Urea
   d. DAP

Answer the following

1. Complex fertilizers
2. Fertilizer mixtures
3. Green manuring
4. Time of application of manures and fertilizers
5. Integrated nutrient management.

15

Harvesting and Post Harvest Technology

Harvesting: "Removal of entire plant or economic parts after maturity from the field" is called *harvesting.* It includes the operation of cutting, picking, plucking or digging or a combination of these for removing the useful part or economic part from the plants / crops. The portion of the stem that is left in the field after harvest is called as *stubble*. The economic product may be *grain, seed, leaf, root* or *entire plant.*

Harvest Index: (H.I): It is the ratio of the economic yield to the total biological yield expressed as percentage.

H.I = Economic yield / Biological yield x 100

Time of Harvesting: If the crop is harvested early, the produce contains high moisture and more immature ill filled and shriveled grains. High moisture leads to pest attack and reduction in germination percentage and impairs the grain quality.

Late harvesting results in shattering of grains, germination even before harvesting during rainy season and breakage during processing.

Losses due to later harvesting	Percentage of loss
Harvesting at physiological maturity	0.71
Harvesting at harvest maturity	3.50
Harvesting one week after maturity	5.63
Harvesting two week after maturity	8.64
Harvesting three week after maturity	14.70
Harvesting four week after maturity	16.40

Hence, harvesting at correct time is essential to get good quality grains and higher yield.

Time of harvesting can be assessed by

i) Calculating the growing degree days (GDD)

ii) Assessing maturity from the duration of crop

iii) Growing Degree Days: A degree day or a heat unit is the mean temperature above base temperature,

$$GDD = \sum_{i=i}^{n} \frac{[T max + T min]}{2} - Tb$$

where Tmax is maximum temperature, Tmin is minimum temperature., Tb is the lowest temperature at which there is no growth (base temperature) For example - base temperature of rice, maize and cumbu is 10°C whereas it is 4.5°C for wheat. Degree days are useful for predicting the time of harvest by calendaring the required photo thermal units (PTU) to complete each growth stage of the crop.

$$PTU = \sum_{i=i}^{n}$$ (GDD x Length of day for long day plant or night for short day plant)

ii) **Assessing Maturity**: Crops can be harvested by assessing the maturity i.e. at physiological maturity or at harvest maturity.

a) **Physiological maturity** refers to a development stage after which no further increases in dry matter occurs in the economic part. Crop is considered to be at physiological maturity when the translocation of photosynthesis to the economic part is stopped.

b) **Harvest maturity** generally occurs seven days after physiological maturity. The important process during this period is loss of moisture from the plants.

External symptoms of physiological maturity

The major symptoms of physiological maturity of some field crops are as follows:

i) *Wheat and Barley* – Complete loss of green colour from the glumes.

ii) *Maize and Sorghum* – Black layer in the placental region of grain

iii) *Pearlmillet* – Appearance of bleached peduncle

iv) *Soybean* – Loss of the green colour from leaves.

v) *Redgram* – Green pods turning brown about 25 days after flowering.

Harvest maturity symptoms

The harvest maturity symptoms of some important crops are as follows :

i) Rice – Hard and yellow coloured grains.

ii) Wheat – Yellowing of spikelets.

iii) Sorghum, Pearl millet, foxtail millet – Yellow coloured ears with hard grains.

iv) Ragi – Brown coloured ears with hard grains

v) Pulses – Brown coloured pods with hard seeds inside the pods.

vi) Groundnut – Inner side of the pods turn dark from light color.

vii) Sugarcane – Leaves turn yellow. Sucrose content is less than 15% and brix reading is more than 18%

viii) Tobacco – leaves slightly turn yellow in colour and specks appear on the leaves.

Criteria for Harvesting of Crops

Crop	Criteria for harvesting
Rice	32 days after flowering Green grains not more than 4-9%
Wheat	About 15% moisture in grain Grain in hard dough stage.
Maize	25 – 30 days after tasselling, Seed moisture content is at 34%
Sorghum	40 days after flowering
Cumbu	28 – 35 days after flowering
Redgram	35 – 40 days after flowering
Black / Green gram	Pod turn brown / black
Rapeseed / mustard	75% of the silique turn yellowSeed moisture at 30%
Sunflower	Back of heads turns to lemon yellow
Groundnut	Yellowing of leaves and shedding Development of purple colour of the testa
Cotton	Bolls fully opened
Jute	50% pod stage (120 – 150 days)
Sugarcane	Brix 18 – 20%Sucrose 15%

Determination of harvesting date is easier for determinate crops and difficult for indeterminate crops because at a given time, the indeterminate plants contain flowers, immature and mature pods or fruits. If the harvesting is delayed for the sake of immature pods, mature pods may shatter. If harvested earlier, yield is less due to several immature pods. This problem can be overcome by,

i). harvesting pods or ears when 75% of them are mature (or)

ii) periodical harvesting or picking of pods

iii) inducing uniform maturity by spraying Paraquat or 2,4-D sodium salt.

In fodder crops, toxins present in the crop, nutritive value, purpose of harvest (whether for stall feeding or for storage) and single or multi cut are also to be considered during harvest. Example - HCN toxin content in sorghum is high upto 30 - 45 DAS.

Methods of Harvesting

Harvesting is done either manually or by mechanical means.

Manual: Sickle is the important tool used for harvesting. The sickle has to be sharp, curved and serrated for efficient harvesting. Knife is used for harvesting of plants with thick and woody stems. Nowadays improved type of sickle is available which reduce the druggery of harvesting labourers.

Mechanical: Harvesting with the use of implements or machines

Implements /machinery used for harvest, threshing and drying

For harvesting: Power tiller operator paddy harvester, combine harvester, guntaka etc., are used

i) **Paddy harvester**: It is used for non-lodging varieties. During operation, the nose of the harvester first enters into the standing crop and movement brings the crop between star projections of the wheel, then it cuts the standing crop. Capacity of this machine is one hectare per day. The width of the coverage for one movement will be 0.75 m. The power requirement is 3 H.P. The present cost of the unit will be Rs.30,000 and operation cost is Rs.260/ha including cost of two labourers.

ii) **Combine harvester**: It is possible to harvest and thresh the produce simultaneously using combine harvester. It cuts the crop, separates the grain from straw, cleans it from chaff and dust and stores the grains in the storage tank. The combine harvester reaps 2-9 rows at a time depending on its size and is equipped with 8 to 10 H.P engine. The cutting operation is done by a reciprocating type of cutter bar with a speed of 800-900 strokes per minute. The cut portion is transferred to conveyor belt or plant form with the help of wheel. Threshing cylinders operating at a peripheral speed of 800-1200 stroked per minute are used for threshing. Grain and chaff are separated with the help of blowers.

iii) **Guntaka**: Ground nut is harvested using heavy blade harrows called Guntakas – R.E. Guntaka

For threshing following are used

i) Olpad thresher,
ii) Japanese rotary paddy thresher,
iii) multi crop thresher,
iv) rollers etc.

i) **Olpad** thresher is used for wheat, barley, oats etc. It consists of 20 circular discs each 45 cm in diameter and 3 mm in thickness placed 15 cm apart in three rows run by pair of bullocks over the dried crop spread circularly on the threshing floor.

ii) **Japanese paddy thresher** consists of a threshing drum, driving mechanism and a supporting frame. Main parts are wooden drum with peg-teeth all around its circumference. The diameter of the drum is about 43 cm to 76 cm. The thresher is operated by a single person with the help of a pedal. Threshing of paddy is done by holding the bundle of harvested material against the teeth of revolving drum.

iii) **Multi-crop threshers**: Mechanical thresher commonly used for threshing major cereals, oil seeds and pulses. It is operated by an electric motor or oil engine. These threshers have provision to control concave clearance and threshing drum and blower speed independently so as to reduce grain breakage and improve cleaning. Sunflower and safflower which are difficult to thresh with traditional methods can also be threshed by the multi-crop thresher.

iv) **Rollers** made of stone are used to thresh grains from ears of millets like ragi, sorghum and cumbu. Earheads are spread to a thickness of 20 cm in a circular fashion on a threshing floor and rollers are drawn over it by a pair of bullocks.

Drying is done either by using solar energy or by artificial heating (mechanical drying) of air and circulating it as in driers.

***Storage*:** Though harvesting of crops is seasonal, consumption of the food grains is continuous. The market value of the produce is generally low at harvesting time. Therefore, there is necessity to store the produce for different periods. The different categories of agriculture produce needing storage are food grains, pulses, oilseeds and seeds.

Post Harvest Processing

Post harvest processing encompasses an array of handling and processing system from the stage of maturation till consumption of the produce and includes threshing, cleaning, grading, drying, parboiling, curing, milling, preservation, storage, processing, packaging, transportation, marketing and consumption systems.

The most important factor deciding the storability of the produce is moisture content of the produce. High moisture content invites pest and disease and induce pre-germination. Moisture content for safe storage of grains of most crops is about 14% (raw rice), 15% for parboiled rice, 12% for wheat, barley, other millets and pulses, 10% for coriander, chillies and 6% for groundnut, rapeseed and mustard.

Objectives of post harvest processing technologies

1. To minimize post harvest losses this is around 10-25% in cereals and 20-30% in perishables.
2. To get good quality products
3. To get maximum quantity of materials by way of proper PHT.
4. To get value added products by way of processing.
5. For proper utilization of water from food industries.
6. To create employment opportunities.
7. To eliminate or minimize the pollution.

Principles involved in the post harvest processing of rice

i) Threshing: Involves the detachment of grains from the panicle

ii) Drying: Reduction of 12 – 14% or 8% by evaporation. i.e., it involves heat and mass transfer operations simultaneously.

iii) Parboiling: Is a hydrothermal treatment followed by drying before milling for the production of milled parboiled grain. The most important change during parboiling is the gelatinization of starch and disintegration of protein bodies in the endosperm.

iv) Milling: Refers to the size reduction and separation operations used for processing of food grains into edible form by removing and separating the inedible and undesirable portions from them, Milling may involve cleaning/separating husk (dehusking), sorting, whitening, polishing, grinding etc.,

v) Storage: Proper storage in storage structures is necessary to prevent the grains from storage pest and to maintain the quality of seeds.

Post Harvest Technology

The quantitative losses encountered at various stages are 1 to 3%, during harvest, 2 to 6% during threshing, 1 to 5% during drying 2 to 7% during handling 2 to 10% during milling and 2 to 6% during handling 2 to 10% during milling and 2 to 6% during storage. To overcome these losses the following improved practices can be adopted.

i) **Harvesting**: Paddy if not harvested at the optimum time, results in loss of quality and quantity. To reduce these losses, machines like combines and reapers are being introduced to harvest paddy at an appropriate stage.

ii) **Threshing**: Threshing, done by bullocks, tractors and by hand, result in poor drying, storage and milling. The multicrop threshers have been developed to reduce these losses.

iii) **Transporting**: Poor transport facilities result in losses to the farmers, millers, and eventually food grain to the country, sometimes as much as 2 to 3 per cent. Good transport facilities should be used to minimize these losses.

iv) **Drying:** Sun drying methods cause more breakage of grain than other factor, resulting in low head yields and low milling yields. Moist paddy in storage deteriorates rapidly. With the introduction of heated air dryers, the losses can be reduced considerably.

v) **Storage:** Uncleaned wet paddy accounts for the largest losses during storage. This is followed by losses due to rodents, birds, mould, fungus, insects and pilferage. These losses can be minimized by storing in good storage structures.

Moisture content of grains for safe storage

Crop	Moisture content (%)
Paddy and raw rice	14
Parboiled rice	15
Wheat, barley, maize, millets, and pulses	12
Coriander, chilli, fenugreek	10
Groundnut pods, rape and mustard	6

vi) **Parboiling**: It is done by soaking the grain in large concrete tanks and steaming it in small kettles or Soaking and steaming grain in large metal tanks with a boiler. The old traditional methods of parboiling incur physical losses and excessive cost of operation. Modern parboiling technologies have been developed and widely accepted by millers with good success leading reduction in losses.

vii) **Milling**: Traditional milling equipment has the lowest milling recovery. With modernization of rice milling industry and by replacing hullers with modern mills, the milling losses can be reduced.

The qualitative losses like change in colour, odour, vitamins, texture are due to over exposure to sun, fungal growth and insect attack. These losses can also be controlled by proper handling, drying and storage after harvest.

Questions

I. Fill in the blanks

1. Removal of entire plant or economic parts after maturity from the field is called ________________.
2. The ratio of the economic yield to the total biological yield expressed as percentage is called ______________.
3. Moisture content for safe storage of grains of most crops is about ____%
4. A development stage after which no further increases in dry matter occurs in the economic part is ________________.

5. The portion of the stem left in the field after harvest is called as ________.

II. Choose the correct answer

6. Early harvest of the crop results in
   a. Shattering of grains
   b. Breakage during processing.
   c. Ill filled and shrivelled grains
   d. Good quality grains
7. Late harvest of the crops results in
   a. High moisture b. Ill filled grains
   c. Shrivelled grains d. Shattering of grains
8. Determination of harvesting date is easier for
   a. Determinate crops b. Indeterminate crops
   c. Perennial crops d. None
9. Cotton can be harvested when bolls are
   a. Fully opened b. Young
   c. Half opened d. None of these
10. The important process that occurs during harvest maturity is
   a. Loss of moisture from the plants
   b. Grain filling
   c. Translocation of photosynthates to the grain
   d. Physiological maturity

III. Answer the following

1. Objectives of post-harvest processing
2. Parboiling in paddy
3. Mechanical harvesting
4. Time of harvesting
5. Paddy harvester.

Selected References

Biwas, T.D. and S.K.Mukherjee. 1994. Text Book of Soil Science, Tata Mc Graw Hill Publishing Company Ltd., New Delhi.

Brady, N.C. 1990. The Nature and Properties of Soils, Mac Millan Publishing Co., New York, USA.

Cheema, S.S., Dhaliwal, B.K. and T.S.Sahota. 2000. Theory and Digest Agronomy, Kalyani Publishers, New Delhi.

Dahama, A.K. 1996. Orgnaic Farming for Sustainable Agriculture, Agri. Botanical Publishers, Bikaner.

DAS, P.C. 2000. Manures and Fertilizers. Kalyani Publishers, New Delhi.

Gopal Chandra De. 1997. Fundamentals of Agronomy, Oxford and IBH Publishing Co. Pvt. Ltd., New Delhi.

ICAR. 1996. Handbook of Agriculture, Indian Council of Agricultural Research, New Delhi.

Mandal, R.C. and P.K. Jana. 1998. Water Resource Utilization and Micro-irrigation; Sprinkler and Drip System, Kalyani Publishers, Ludhiana.

Michael, A.M. 1978. Irrigation – Theory and Practice, Vikas Publishing House Pvt., Co. New Delhi.

Morachan, Y.B. 1986. Crop Production and Management, Oxford and IBH Pub. Co.New Delhi.

Palaniappan, SP. 1985. Cropping systems in the Tropics – Principles and Management, Wiley Eastern Limited and Tamil Nadu Agricultural University, Coimbatore.

Randhawa, N.S. 1980. A History of Agriculture in India, Vols. I & II Indian Council of Agricultural Research, New Delhi.

Rao, V.S. 1983. Principles of Weed Science. Oxford and IBH, Pub. Co. New Delhi.

Reddy, S.R. 1999. Principles of Agronomy. Kalyani Publishers, New Delhi.

Sankaran, S and V.T. Subbiah Mudaliar. 1997. Principles of Agronomy. The Bangalore Printing and Publishing co. Ltd., Bangalore.

Singh, S.S. 1999. Principles and Practices of Agronomy, Kalyani Publishers, New Delhi.

Somasundaram, E., Annadurai, K., Manoharan, M.L., Kavimani, R and R. Vijayalakshmi. 2000. Hand book on Rice Production Technology, Golden Net printers, Trichirappalli.

Somasundaram, E., Mohamed Amanullah, M. Ashok Raja, N. and D. Udhyanandhini (2015). "A quick guide to Organic Farming", TNAU Publications, Coimbatore.

Subbian, P., Annadurai, K and SP. Palaniappan. 2000. Agriculture Facts and Figures, Kalyani Publishers, New Delhi.

Thakur, C. 1980. Scientific Crop Production Vol.I, Metropolitan Book Co. Pvt. Ltd., New Delhi.

Thakur, C. 1981. Scientific Crop Production Vol. II, Metropolitan Book Co. Pvt. Ltd., New Delhi.

Yadav, J.S.P. and G.B. Singh. 2000. (Eds.,) Natural Resource Management for Agricultural Production in India, Indian Council of Agricultural Research, New Delhi.

Yellamanda Reddy, T and Sankara Reddi, G.H. 1999. Principles of Agronomy, Kalyani Publishers, New Delhi.

17

Annexures

ANNEXURE – I

Units Related to Crop Production

1. Area

a) Inch: It is equal to 2.54 cm.

b) Feet: It is equal to 30.48 cm or 0.305 m

c) Cent: It is an unit of measurement of an area of land which is 1/100 of an acre. This is equal to 40 m^2 or 435.6 sq. ft.

d) Acre : An unit of measurement of an area to 100 cents (4000 m^2). It is also equal to 2/5th of hectare.

e) Acre : 1/100th of a hectare.

f) Hectare: It refer to an area of 10,000 m^2 or 250 cents or 2.5 acres/h 100 kuzhi = 1 maas or kaani ; 3 maa = 1 acre; 20 maa = 6.67 acres = 1 veli

2. Weight

a) Pound: It is a unit of weight equal to 454 grams.

b) Kilogram: It is unit of weight equal to 1000 grams or 2.203 pounds.

c) Quintal: This refers to an unit of weight equal to 100 kg. or 0.1 tonne.

d) Bale: 177.8 kg. (cotton lint)

e) Tonne: Unit of weight equal to 1000 kgs or 10 quintols.

f) Million tonnes: 10,00,000 tonnes

3. Volume

a) Milli litre (ml): It is an unit of volume equal to 1/1000th of a litre. 1 ml = 1 cc = 1cu. mm.

b) Litre: It is a volume equal to 1000 ml or 1000 cc.

c) Cubic metre: It is a volume equal to 1000 litres.

d) Cubic foot: It is a volume equal to 28.32 litres

e) TMC: Thousand metric cubic feet

f) Gallon: 4.546 litres.

ANNEXURE – II

A. Area, Production and Productivity of Major crops in India (2011-12)

S.No.	Crop	Area (m/ha)	Production (mt)	Productivity (kg/ha)
1.	Rice	43.97	104.32	2372
2.	Wheat	29.90	93.90	3140
3.	Maize	8.71	21.57	2476
4.	Jowar	6.32	6.03	954
5.	Pearl millet	8.69	10.05	1156
6.	Pulses	24.78	17.21	694
7.	Oilseeds	26.44	30.01	1135
8.	Groundnut	5.31	6.93	1305
9.	Rapeseed and mustard	5.92	6.78	1145
10.	Soyabean	10.18	12.28	1207
11.	Sunflower	0.72	0.50	692
12.	Cotton	12.18	35.20	491
13.	Jute & Mesta	0.91	11.57	2283
14.	Sugarcane	5.09	357.67	70317

B. Area, Production and Productivity of Major crops in Tamilnadu (2011-12)

S. No.	Crop	Area (lakh ha)	Production (lakh tonnes)	Productivity (kg/ha)
1.	Rice	19.0	74.5	3918
2.	Sorghum	1.98	2.52	1277
3.	Bajra	0.46	1.14	2452
4.	Ragi	0.83	2.25	2715
5.	Maize	2.80	16.95	6042
6.	Small Millets	0.30	0.36	1210
7.	Redgram	0.36	0.31	870
8.	Blackgram	3.08	1.79	580
9.	Green gram	1.64	0.85	519
10.	Horsegram	0.69	0.37	539
11.	Bengal gram	0.08	0.05	641
12.	Groundnut	3.86	10.61	2750
13.	Gingelly	0.43	0.26	612

14.	Castor	0.06	0.02	309
15.	Coconut (no. of nuts)	4.19	0.62	14799
16.	Cotton (Bales of 170 kg lint)	1.36	3.82	481
17.	Sugarcane (t/ha)	3.46	389.75	112
18.	Tobacco	0.03	0.04	1527
19.	Chilli	0.56	0.25	436

ANNEXURE – III

List of Crops – Common and Botanical Names

Cereals

1. Rice - *Oryza sativa* Linn.
2. Wheat - *Triticum aestivum* L. *Triticum sativum*, Lamk.
3. Maize - *Zea mays* Linn.
4. Rye - *Secale cereale* Linn.
5. Oat - *Avena sativa* Linn.
6. Barley - *Hordeum vulgare* Linn.
7. Sorghum (Jowar) - *Sorghum bicolor* Pers.
8. Pearl millet (Bajra) - *Pennisetum glaucum* Linn.
9. Finger millet (Ragi) - *Eleusine coracana* Gaertn.
10. Barnyard millet (Kutiraivali) - *Echinochloa frumentacea* Roxb.
11. Italian millet (Thenai) - *Setaria italica*. Linn.
12. Kodo millet (Varagu) - *Paspalum scrobiculatum*. Linn.
13. Common millet (Panivaragu) - *Panicum millaceum* Linn.
14. Little millet (Samai) - *Panicum milleare* Linn.

Pulses

1. Blackgram (Urd) - *Vigna mungo* var, *radiatus* Linn.
2. Chickling vetch (Khesari) - *Lathyrus sativus* Linn.
3. Chickepea (Gram) - *Cicer arietinum* Linn.
4. Cowpea - *Vigna sinensis* Savi
5. Greengram (Mung or Moong) - *Vigna radiatus* Roxb.
6. Horsegram (Kulthi) - *Macrotyloma uniflorum* Linn.
7. Lentil - *Lens esculenta* Moench
8. Moth bean - *Phaseolus aconitifolia* Linn.
9. Peas - *Pisum sativum* Linn.
10. Pigeonpea (Arhar/Tur) - *Cajanas cajan* Millsp. (*Cajanus indicus*)
11. Pillipesera - *Phaseolus trilobus*
12. Soybean - *Glycine max*. Linn. Merr.

Oilseeds

1. Black mustard - *Brassica nigra* Linn. Koch.
2. Castor *Ricinus communis* Linn.
3. Coconut *Cocus nucifera* Linn.

4. Groundnut/peanut - *Arachis hypogaea* Linn.
5. Indian mustard or rai - *Brassica Cass* Linn.
6. Indian rape or toria - *Brassica napeustris* Linn., var, napus
7. Niger - *Guizotia abyssinica* Cass
8. Linseed - *Linum usitatissimum* Linn.
9. Safflower - *Carthamus tinctorious* Linn.
10. Sesame/Gingelly/Til - *Sesamum indicum* Linn.
11. Sunflower - *Helianthus annuus* Linn.
12. White mustard - *Brassica alba* Linn.
13. Oilpalm - *Elaeis guinensis*

Fibre Crops

1. Cotton - *Gossypium* spp.
2. Jute - *Corchorus* spp.
3. Mesta - *Hibiscus cannabinus* Linn.
4. Sunnhemp - *Crotalaria juncea* Linn.

Fumitories, Masticatories

1. Tobacco (Desi) - *Nicotiana tabacum*
2. Tobacco (Calcutta) - *Nicotiana rustica*

Sugars and Starches

1. Pine apple - *Ananas sativa* Schutt.
2. Potato - *Solanum tuberosum* Linn.
3. Sugar beet - *Beta vulgaris* Linn.
4. Sugarcane - *Saccharum officinarum* Linn.
5. Sweet potato - *Ipomea batatus* Lann.
6. Tapioca - *Manihot esculenta crantz.*

Spices and Condiments

1. Black pepper - *Piper nigrum* L.
2. Cardamon - *Elettaria cardamomum Matora.*
3. Coriander - *Coriandrum sativum* Linn.
4. Garlic - *Allium sativum* Linn.
5. Ginger - *Zingiber officinale* Rose
6. Onion - *Allium cepa* Linn.
7. Red pepper (Chilli) - *Capsicum annuum* Linn.
8. Turmeric - *Curcuma longa* Linn.

Forage Grasses

1. Buffel grass, Anjan - *Cenchrus ciliaris*
2. Dallis grass - *Paspalum dilatatum* Poir.
3. Dinanath grass - *Pennisetum*
4. Guinea grass - *Panicum maximum* Jacq.
5. Marvel grass - *Dicanthium annulatum* (Forsk.)
6. Napier or Elephat grass - *Pennisetum purpureum* Schum.
7. Pangola grass - *Digitaria decumbens* stent.
8. Para grass - *Brachiaria mutica*
9. Sudan grass - *Sorghum sudanense* Stapf.
10. Teosinte - *Euchlaena mexicana* Schrad.
11. Blue panicum - *Panicum antidotale* Retz.

Forage Legumes

1. Berseem/ Egyptian Clover - *Trifolium alexandrinum* Linn.
2. Centrosema - *Centrosema pubescens*
3. Gaur / Cluster bean - *Cyamopsis tetragonoloba* Taub
4. Lucerne / Alfalfa - *Medicago sativa* Linn.
5. Siratro - *Macroptlium atropurpureum*
6. Velvet Bean - *Mucuna cochinchinensis* Brot.

Plantation Crops

1. Banana - *Musa paradisiaca* L.
2. Areca Palm - *Areca catechu* Linn.
3. Arrowroot - *Maranta arundinacea* L.
4. Cocoa - *Theobroma cocoa* Linn.
5. Coconut - *Cocos nucifera* Linn.
6. Coffee - *Coffea arabica* Linn.
7. Tea - *Camellia theasinesis* O. Ktze.

Green Manure Crops

1. Daincha - *Sesbania aculeata* Poir.
2. Sunnhemp - *Crotolaria juncea Linn.* (Sanappai)
3. Manila agathi - *Sesbania rostrata*
4. Sithagathi - *Sesbania sesban*

Vegetables

1. Ash Gourd - *Beniacasa cerifera* Savi.
2. Bitter gourd - *Momordica charantia* Linn.
3. Bottle gourd - *Lagenaria leucantha* Rusby.
4. Brinjal - *Solanum melongena* Linn.
5. Broad bean - *Vicia faba* Linn.
6. Cabbage - *Brassica oleracea* var. capitata Linn.
7. Chinese cabbage - *B.pekinensis* (Lour) Rupr.
8. Carrot - *Daucus carota* Linn.
9. Cauliflower - *Brassica oleracea* var. botrytis Linn.
10. Colocasia -. *Colocasia esculenta* (L). Schott.
11. Cucumber - *Cucumis sativus* Linn.
12. Double bean - *Phaseolus lunatus* Linn.
13. Elephant ear / edible arum - *Colocasia antiquorum* Schott.
14. Elephant foot / yam - *Amorphophallus campanulatus* Bheme.
15. French bean - *Phaseolus vulgaris* Linn.
16. Knolkhol - *Brassica oleracea* var. Caulorapa Pasq.
17. Lesser yam - *Dioscorea alata* L.
18. Lettuce - *Lactuca sativa* Linn.
19. Musk Melon - *Cucumis melo* Linn.
20. Pointed gourd / Parwal - *Trichosanthes diora* Roxb.
21. Pumpkin - *Cucurbita moschata* Dutch.
22. Radish - *Raphanus sativus* Linn.
23. Bhendi - *Abelmoschus esculentus* Linn.
24. Red pumpkin - *Cucurbita maxima* Duch.
25. Ridge gourd - *Luffa acutangula* Roxb.
26. Spinach - *Spinacia oleracea* Linn.
27. Snake gourd - *Trichosanthes anguina*. Linn.
28. Tomato - *Lycopersicum esculentus* Mill.
29. Turnip - *Brassica campestris* var. rapa Linn.
30. Water melon - *Citrullus vulgaris* schrad.
31. Yam - *Dioscorea esculenta* L.

Medicinal Crops

1. Aloe - *Aloe vera*
2. Ashwagantha - *Withania somnifera* Dunai.
3. Belladonna - *Atropa belladona* Linn.

4. Bishop's weed - *Ammi visnaga* Linn.
5. Bringaraj - *Eclipta alba*
6. Cinchona - *Cinchona sp.*
7. Coleus - *Coleus forskholli* Briq .
8. Dioscorea - *Dioscorea bulbifera* Linn.
9. Duboisia - *Duboisia myoporoides* Brown.
10. Glory Lily - *Gloriosa superba* Linn.
11. Ipecae - *Cephaelis ipecacuanha* Linn.
12. Long pepper - *Piper longum* Linn.
13. Opium poppy - *Papav somniferum*
14. Prime rose - *Oenothera lamarekiana* Linn.
15. Roselle - *Hibiscus sabdariffa* Linn.
16. Sarpagandha - *Rauvalfia serpentine* Benth.
17. Senna - *Cassia angustifolia* Vahl.
18. Sweet Flag - *Acorus calamus* Linn.
19. Valeriana - *Valeriana wallaichii*

Aromatic Crops

1. Ambrettee - *Abelmoschus moschatus* Medic.
2. Celery - *Apium graveolens* Linn.
3. Citronella - *Cymbopogon winterianus* Jowitt.
4. Geranium - *Pelargonium graveolens*
5. Jasmine - *Jasminum grantiflorum*
6. Khus - *Vetiveria zizanoids*
7. Lavender - *Lavendula sp.* Linn
8. Lemon grass - *Cymbopogon flexuosus* Stapf.
9. Mint - *Mint sp.*
10. Palmarosa - *Cymbopogon martini*
11. Patchouli - *Pogostemon cablin* Benth.
12. Sandal wood - *Santalum album*
13. Sacred Basil (Tulsi) - *Ocimum sanctum* Linn.
14. Tuberose - *Polianthus tuberosa* Linn.

Other Economic Crops

1. Annatto - *Bixa orellana*
2. Camphor Basil - *Ocimum kilimandscharicum*
3. Henna - *Lawsonia inermis* Linn.
4. Pyrethrum - *Chrysanthemum cinerarifolium*

ANNEXURE IV

List of Major Weeds in the World

Common Name	Scientific Name	Growth habitat and kind of plant
Smooth pig weed	*Amaranthus hybridus*	A-B
Spiny amaranth	*Amaranthus spinosus*	A-B
Wild oat	*Avena fatua*	A-G
Common lambsquarters	*Chenopodium album*	A-B
Field bind weed	*Convolvulus arvensis*	P-B
Bermuda grass	*Cyanodon dactylon*	P-G
Yellow nut sedge	*Cyprus esculentus*	P-S
Purple nut sedge	*Cyprus rotundus*	P-S
Crab grass	*Digitaria sanguinalis*	A-G
Jungle rice	*Echinochloa colonum*	A-G
Barnyard grass	*Echinochloa crusgalli*	A-G
Water hyacinth	*Eichhornia crassipes*	P-B
Goose grass	*Eleusine indica*	A-G
Cogon grass	*Imperata cylindrical*	P-G
Sour paspalum	*Paspalum conjugatum*	P-G
Common purslane	*Portulaca oleracea*	A-B
Itch grass	*Rottboellia exaltata*	A-G
Johnson grass	*Sorghum halepense*	P-G

A - Annual

P - Perennial

B - Broad Leaved Weed

G - Grasses

S - Sedges

ANNEXURE V

Twenty Six Most Common Weeds in Crop Fields of India

Monocot species	Dicot species
Annuals	
Barnyard grass *(Echinochloa crusgalli)*	Goat weed *(Ageratum conyzoids)*
Crabgrass *(Digitaria sp)*	Pig weed *(Amaranthus sp)*
Foxtail *(Setaria sp.)*	Black jack *(Bidens pilosa)*
Sandbur *(Cenchrus sp)*	Cox comb *(Celosia argentia)*
Wild oat *(Avena fatua)*	Lambsquarters *(Chenopodium album)*
Goose grass *(Eleusine indica)*	Wild carrot *(Parthenium hysterophorus)*
Torpedo grass *(Panicum repens)*	Common purslane *(Portulaca oleracea)*
Canary grass *(Phalaris minor)*	Horse purslane *(Trianthema portulacastrum)*
Crosfoot grass *(Dactyloctenium aegyptium)*	
Perennials	
Bermuda grass *(Cyanodon dactylon)*	Canada thistle *(Circium arvense)*
Thatch grass *(Imperata cylindrical)*	Day flower *(Commelina benghalensis)*
Johnson grass *(Sorghum halepense)*	Field bind *(Convolvulus arvensis)*
Quack grass *(Agropyron repens)*	White horse nettle *(Solanum elaeagnifolium)*
Nut grass *(Cyprus rotundus)*	

ANNEXURE VI

Contribution of Agriculture to National Income

Year	Percentage Contribution of Agriculture and Allied Activities to National Income
1950-1951	56.1
1960-1961	51.2
1970-1971	50.6
1980-1981	42.0
1984-1985	36.9
1989-1990	30.0
1990-1991	29.0
1999-2000	25.5
2011-2012	17.0
2013-2014	17.9
2014-2015	17.0

India ranks second worldwide in farm output. Agriculture and allied sectors like forestry, logging and fishing accounted for 17% of the GDP and employed 51% of the total workforce in 2012. As Indian economy has diversified and grown, agriculture's contribution to GDP has steadily declined from 1951 to 2011, yet it is still the largest employment source and a significant piece of the overall socio-economic development of India

ANNEXURE VII

Food Grains Production in India - A Scenario

Year	Food grain production (million tones)
1950-51	50.82
1960-61	82.02
1970-71	108.42
1980-81	129.59
1983-84	152.40
1984-85	145.50
1985-86	150.40
1986-87	143.40
1987-88	140.40
1988-89	169.90
1989-90	171.10
1990-91	176.39
1993-94	175.91
1996-97	199.30
1997-98	192.43
1998-99	195.25
2000-01	196.80
2001-02	213.00
2002-03	175.00
2003-04	213.00
2004-05	198.00
2005-06	209.00
2006-07	217.00
2007-08	213.00
2008-09	234.00
2009-10	218.00
2010-11	245.00
2011-12	259.00
2012-13	255.00
2013-14	248.76
2014-15	252.02
2015-16*	252.23

* As per third advance estimates.

□□□